Laser-Material Interactions at Micro/Nanoscales

MATERIALS RESEARCH SOCIETY
SYMPOSIUM PROCEEDINGS VOLUME 1365

Laser-Material Interactions at Micro/Nanoscales

Symposium held April 25–29, 2011, San Francisco, California, U.S.A.

EDITORS

Yongfeng Lu
University of Nebraska–Lincoln
Lincoln, Nebraska, U.S.A.

Craig B. Arnold
Princeton University
Princeton, New Jersey, U.S.A.

Costas P. Grigoropoulos
University of California–Berkeley
Berkeley, California, U.S.A.

Michael Stuke
Max-Planck-Institute for Biophysical Chemistry
Göttingen, Germany

Steven M. Yalisove
University of Michigan
Ann Arbor, Michigan, U.S.A.

MRS

Materials Research Society
Warrendale, Pennsylvania

CAMBRIDGE
UNIVERSITY PRESS

CAMBRIDGE
UNIVERSITY PRESS

Shaftesbury Road, Cambridge CB2 8EA, United Kingdom

One Liberty Plaza, 20th Floor, New York, NY 10006, USA

477 Williamstown Road, Port Melbourne, VIC 3207, Australia

314–321, 3rd Floor, Plot 3, Splendor Forum, Jasola District Centre, New Delhi – 110025, India

103 Penang Road, #05–06/07, Visioncrest Commercial, Singapore 238467

Cambridge University Press is part of Cambridge University Press & Assessment, a department of the University of Cambridge.

We share the University's mission to contribute to society through the pursuit of education, learning and research at the highest international levels of excellence.

www.cambridge.org
Information on this title: www.cambridge.org/9781605113425

First published 2012

CODEN: MRSPDH

A catalogue record for this publication is available from the British Library

ISBN 978-1-605-11342-5 Hardback

CONTENTS

*Invited Paper

*Invited Paper

PREFACE

This volume includes a collection of papers presented in Symposium TT: *Laser-Material Interactions at Micro/Nanoscales*, at the Spring 2011 Materials Research Society Meeting held April 2011 in San Francisco, California. Laser-material interactions are of fundamental importance in wide range of materials-related research and technology areas including green energy, photonics, electronics, environmental studies, biomedical imaging, medical treatment, and optical spectroscopy. This symposium provided an interdisciplinary forum for scientists and engineers from different fields to discuss the physical, chemical, thermal, and mechanical phenomena that occur during laser-material interactions at micro and nanoscales. Research addressing new materials, processes, structures, and surfaces synthesized by these methods for emerging fields was discussed. The symposium was well attended and received a large number of abstract submissions in the areas of ultrafast laser processing, laser ablation and deposition, process controls, nanomaterials, surface modification, laser materials interactions, polymerization, lithography, and novel approaches in laser processing.

The editors of this proceedings volume would like to express their sincere appreciation to all participants of this symposium who together helped to create an interactive environment in which to share scientific and technological developments in the field. We especially thank the invited speakers and session chairs and are particularly grateful to those contributors who provided full papers for this issue. In addition, we thank the staff at the Materials Research Society who helped to organize this conference and the symposium assistants who helped keep the sessions running on time.

Yongfeng Lu
Craig B. Arnold
Costas P. Grigoropoulos
Michael Stuke
Steven M. Yalisove

October 2011

MATERIALS RESEARCH SOCIETY SYMPOSIUM PROCEEDINGS

Volume 1321 — Amorphous and Polycrystalline Thin-Film Silicon Science and Technology—2011, B. Yan, S. Higashi, C.C. Tsai, Q. Wang, H. Gleskova, 2011, ISBN 978-1-60511-298-5
Volume 1322 — Third-Generation and Emerging Solar-Cell Technologies, Y. Lu, J.M. Merrill, M.T. Lusk, S. Bailey, A. Franceschetti, 2011, ISBN 978-1-60511-299-2
Volume 1323 — Advanced Materials Processing for Scalable Solar-Cell Manufacturing, L. Tsakalakos, H. Ji, B. Ren, 2011, ISBN 978-1-60511-300-5
Volume 1324 — Compound Semiconductors for Energy Applications and Environmental Sustainability—2011, F. Shahedipour-Sandvik, L.D. Bell, K. Jones, B. Simpkins, D. Schaadt, M. Contreras, 2011, ISBN 978-1-60511-301-2
Volume 1325 — Energy Harvesting—Recent Advances in Materials, Devices and Applications, R. Venkatasubramanian, H. Liang, H. Radousky, J. Poon, 2011, ISBN 978-1-60511-302-9
Volume 1326E — Renewable Fuels and Nanotechnology, H. Idriss, 2011, ISBN 978-1-60511-303-6
Volume 1327 — Complex Oxide Materials for Emerging Energy Technologies, J.D. Perkins, A. Ohtomo, H.N. Lee, G. Herranz, 2011, ISBN 978-1-60511-304-3
Volume 1328E — Electrochromic Materials and Devices, M. Bendikov, D. Gillaspie, T. Richardson, 2011, ISBN 978-1-60511-305-0
Volume 1329E — Nanoscale Heat Transfer—Thermoelectrics, Thermophotovoltaics and Emerging Thermal Devices, K. Nielsch, S.F. Fischer, B.J.H. Stadler, T. Kamins, 2011, ISBN 978-1-60511-306-7
Volume 1330E — Protons in Solids, V. Peterson, 2011, ISBN 978-1-60511-307-4
Volume 1331E — Frontiers of Solid-State Ionics, K. Kita, 2011, ISBN 978-1-60511-308-1
Volume 1332E — Interfacial Phenomena and *In-Situ* Techniques for Electrochemical Energy Storage and Conversion, H. Li, 2011, ISBN 978-1-60511-309-8
Volume 1333E — Nanostructured Materials for Energy Storage, J. Lemmon, 2011, ISBN 978-1-60511-310-4
Volume 1334E — Recent Developments in Materials for Hydrogen Storage and Carbon-Capture Technologies, R. Zidan, 2011, ISBN 978-1-60511-311-1
Volume 1335 — Materials, Processes, and Reliability for Advanced Interconnects for Micro- and Nanoelectronics—2011, M. Baklanov, G. Dubois, C. Dussarrant, T. Kokubo, S. Ogawa, 2011, ISBN 978-1-60511-312-8
Volume 1336E — Interface Engineering for Post-CMOS Emerging Channel Materials, Y. Kamata, 2011, ISBN 978-1-60511-313-5
Volume 1337 — New Functional Materials and Emerging Device Architectures for Nonvolatile Memories, D. Wouters, O. Auciello, P. Dimitrakis, Y. Fujisaki, E. Tokumitsu, 2011, ISBN 978-1-60511-314-2
Volume 1338E — Phase-Change Materials for Memory and Reconfigurable Electronics Applications, B.-K. Cheong, P. Fons, B.J. Kooi, B.-S. Lee, R. Zhao, 2011, ISBN 978-1-60511-315-9
Volume 1339E — Plasma-Assisted Materials Processing and Synthesis, J.L. Endrino, A. Anders, J. Andersson, D. Horwat, M. Vinnichenko, 2011, ISBN 978-1-60511-316-6
Volume 1340E — High-Speed and Large-Area Printing of Micro/Nanostructures and Devices, T. Sekitani, 2011, ISBN 978-1-60511-317-3
Volume 1341 — Nuclear Radiation Detection Materials—2011, A. Burger, M. Fiederle, L. Franks, K. Lynn, D.L. Perry, K. Yasuda, 2011, ISBN 978-1-60511-318-0
Volume 1342 — Rare-Earth Doping of Advanced Materials for Photonic Applications—2011, V. Dierolf, Y. Fujiwara, T. Gregorkiewicz, W.M. Jadwisienczak, 2011, ISBN 978-1-60511-319-7
Volume 1343E — Recent Progress in Metamaterials and Plasmonics, G.J. Brown, J. Pendry, D. Smith, Y. Lu, N.X. Fang, 2011, ISBN 978-1-60511-320-3
Volume 1344 — Functional Two-Dimensional Layered Materials—From Graphene to Topological Insulators, A.A. Balandin, A. Geim, J. Huang, D. Li, 2011, ISBN 978-1-60511-321-0
Volume 1345E — Nanoscale Electromechanics of Inorganic, Macromolecular and Biological Systems, J. Li, S.V. Kalinin, M.-F. Yu, P.S. Weiss, 2011, ISBN 978-1-60511-322-7
Volume 1346 — Micro- and Nanofluidic Systems for Materials Synthesis, Device Assembly and Bioanalysis—2011, R. Fan, J. Fu, J. Qin, A. Radenovic, 2011, ISBN 978-1-60511-323-4
Volume 1347E — Nanoscale Heat Transport—From Fundamentals to Devices, A. McGaughey, M. Su, S. Putnam, J. Shiomi, 2011, ISBN 978-1-60511-324-1

MATERIALS RESEARCH SOCIETY SYMPOSIUM PROCEEDINGS

Volume 1348E — Hybrid Interfaces and Devices, D.S. Ginley, N.R. Armstrong, G. Frey, R.T. Collins, 2011, ISBN 978-1-60511-325-8
Volume 1349E — Quantitative Characterization of Nanostructured Materials, A. Kirkland, 2011, ISBN 978-1-60511-326-5
Volume 1350E — Semiconductor Nanowires—From Fundamentals to Applications, V. Schmidt, L.J. Lauhon, T. Fukui, G.T. Wang, M. Björk, 2011, ISBN 978-1-60511-327-2
Volume 1351 — Surfaces and Nanomaterials for Catalysis through *In-Situ* or *Ex-Situ* Studies, F. Tao, M. Salmeron, J.A. Rodriguez, J. Hu, 2011, ISBN 978-1-60511-328-9
Volume 1352 — Titanium Dioxide Nanomaterials, X. Chen, M. Graetzel, C. Li, P.D. Cozzoli, 2011, ISBN 978-1-60511-329-6
Volume 1353E — The Business of Nanotechnology III, L. Merhari, M. Biberger, D. Cruikshank, M. Theelen, 2011, ISBN 978-1-60511-330-2
Volume 1354 — Ion Beams—New Applications from Mesoscale to Nanoscale, J. Baglin, D. ILA, G. Marletta, A. Öztarhan, 2011, ISBN 978-1-60511-331-9
Volume 1355E — Biological Hybrid Materials for Life Sciences, L. Stanciu, S. Andreescu, T. Noguer, B. Liu, 2011, ISBN 978-1-60511-332-6
Volume 1356E — Microbial Life on Surfaces—Biofilm-Material Interactions, N. Abu-Lail, W. Goodson, B.H. Lower, M. Fornalik, R. Lins, 2011, ISBN 978-1-60511-333-3
Volume 1357E — Biomimetic Engineering of Micro- and Nanoparticles, D. Discher, 2011, ISBN 978-1-60511-334-0
Volume 1358E — Organic Bioelectronics and Photonics for Sensing and Regulation, L. Torsi, 2011, ISBN 978-1-60511-335-7
Volume 1359 — Electronic Organic and Inorganic Hybrid Nanomaterials—Synthesis, Device Physics and Their Applications, Z.-L. Zhou, C. Sanchez, M. Popall, J. Pei, 2011, ISBN 978-1-60511-336-4
Volume 1360 — Synthesis and Processing of Organic and Polymeric Materials for Semiconductor Applications, A.B. Padmaperuma, J. Li, C.-C. Wu, J.-B. Xu, N.S. Radu, 2011, ISBN 978-1-60511-337-1
Volume 1361E — Engineering Polymers for Stem-Cell-Fate Regulation and Regenerative Medicine, S. Heilshorn, J.C. Liu, S. Lyu, W. Shen, 2011, ISBN 978-1-60511-338-8
Volume 1362 — Carbon Functional Interfaces, J.A. Garrido, K. Haenen, D. Ho, K.P. Loh, 2011, ISBN 978-1-60511-339-5
Volume 1363 — Fundamental Science of Defects and Microstructure in Advanced Materials for Energy, B.P. Uberuaga, A. El-Azab, G.M. Stocks, 2011, ISBN 978-1-60511-340-1
Volume 1364E — Forum on Materials Education and Evaluation—K-12, Undergraduate, Graduate and Informal, B.M. Olds, D. Steinberg, A. Risbud, 2011, ISBN 978-1-60511-341-8
Volume 1365 — Laser-Material Interactions at Micro/Nanoscales, Y. Lu, C.B. Arnold, C.P. Grigoropoulos, M. Stuke, S.M. Yalisove, 2011, ISBN 978-1-60511-342-5
Volume 1366E — Crystalline Nanoporous Framework Materials—Applications and Technological Feasibility, M.D. Allendorf, K. McCoy, A.A. Talin, S. Kaskel, 2011, ISBN 978-1-60511-343-2
Volume 1367E — Future Directions in High-Temperature Superconductivity—New Materials and Applications, C. Cantoni, A. Ballarino, K. Matsumoto, V. Solovyov, H. Wang, 2011, ISBN 978-1-60511-344-9
Volume 1368 — Multiferroic, Ferroelectric, and Functional Materials, Interfaces and Heterostructures, P. Paruch, E. Tsymbal, M. Rzchowski, T. Tybell, 2011, ISBN 978-1-60511-345-6
Volume 1369E — Computational Studies of Phase Stability and Microstructure Evolution, Z.-K. Liu, 2011, ISBN 978-1-60511-346-3
Volume 1370 — Computational Semiconductor Materials Science, L. Liu, S.-H. Wei, A. Rubio, H. Guo, 2011, ISBN 978-1-60511-347-0

Prior Materials Research Society Symposium Proceedings available by contacting Materials Research Society

Ultra-fast Laser Processing

Mater. Res. Soc. Symp. Proc. Vol. 1365 © 2011 Materials Research Society
DOI: 10.1557/opl.2011.1220

Transient Localized Material Properties Changes by Ultrafast Laser-Pulse Manipulation of Electron Dynamics in Micro/Nano Manufacturing

Xin Li[1], Lan Jiang[1*], Cong Wang[1] and Hai-Lung Tsai[2]
[1]Laser Micro/Nano Fabrication Laboratory, School of Mechanical Engineering, Beijing Institute of Technology, Beijing 100081, People's Republic of China
[2]Laser-Based Manufacturing Laboratory, Department of Mechanical and Aerospace Engineering Missouri University of Science and Technology, Rolla, MO 65409, USA

ABSTRACT

A femtosecond (fs) pulse duration is shorter than many physical/chemical characteristic times, such as the electron-photon relaxation time, which makes it possible to control electron dynamics. This paper reviews our recent progress which proposes to change electron dynamics (selective excitation/ionization) and electron densities/temperatures in materials to control the following properties and processes: 1) the transient (femtosecond-to-picosecond time scale), localized (nanometer-to-micrometer length scale) material properties, 2) the corresponding photon absorption process, and 3) phase change mechanisms, by manipulating fs pulse-train number/delay for high-precision micro/nanoscale manufacturing.

INTRODUCTION

Femtosecond (fs) lasers, with short pulse duration compared with many physical/chemical characteristic times and extremely high irradiance, lead to a new area of ultrafast science and technology, which introduce new phenomena, requiring new experimental and theoretical understanding [1-3]. Recently, developments of optics devices have made fs pulse-train design possible by shaping classic Gauss shape fs pulses to arbitrary shape fs pulses [4], which offer us great opportunity for controlling electron dynamics [5-7]. In this paper, our recent progress is reviewed on the studies of changing electron dynamics to control the following properties and processes by fs pulse-train design for high-precision micro/nanoscale manufacturing [8, 9].

THEORY

For dielectrics, the first-principles calculation (time-dependent density-functional theory, TDDFT) is adopted to study the electron dynamics (selective excitation/ionization) by fs pulse-train irradiation. The time-dependent Kohn-Sham (TDKS) equation for single particle orbitals is applied to describe the motion of electrons [10].

$$i\hbar\frac{\partial}{\partial t}\psi_i(\vec{r},t)=H_{KS}(\vec{r},t)\psi_i(\vec{r},t) \qquad (1)$$

* Author to whom correspondence should be addressed. Electronic address: jianglan@bit.edu.cn.

$$n(\vec{r},t)=\sum_i\left|\psi_i(\vec{r},t)\right|^2 \quad (2)$$

where $n(\vec{r},t)$ is the electron density, $H_{KS}(\vec{r},t)$ is the Kohn-Sham Hamiltonian which is conventional separated in the following way,

$$H_{KS}(\vec{r},t)=\frac{1}{2m}(\vec{p}+\frac{e}{c}\vec{A}(t))^2+V_{ion}(\vec{r},t)+e^2\int d\vec{r}'\frac{n(\vec{r},t)}{\left|\vec{r}-\vec{r}'\right|}+V_{xc}(\vec{r},t) \quad (3)$$

where e is an elementary charge, $\vec{A}(t)$ is a time-dependent spatially-uniform vector potential which is related to the external electric field, $V_{ion}(\vec{r},t)$ is the electron-ion potential and $V_{xc}(\vec{r},t)$ is the exchange-correlation potential.

For metals, a model combined two-temperature model and molecular dynamics simulation is adopted to explore the phase change mechanisms by fs pulse-train design [9]. The two temperature model treats electrons and lattices as two separate subsystems

$$C_e(T_e)\frac{\partial T_e}{\partial t}=\nabla\left[k_e(T_e)\nabla T_e\right]-G(T_e-T_l)+S(z,t)$$

$$C_l(T_l)\frac{\partial T_l}{\partial t}=G(T_e-T_l) \quad (4)$$

where T_e is electron temperature, T_l is lattice temperature, C_e is electron heat capacity, C_l is lattice heat capacity, k_e is electron heat conductivity, G is electron-lattice coupling factor. And the Morse potential is used for interactions among nickel atoms.

$$\Phi(r_{ij})=D\left(e^{-2\alpha(r_{ij}-r_0)}-2e^{-\alpha(r_{ij}-r_0)}\right) \quad (5)$$

where $\Phi(r_{ij})$ is the potential energy of two atoms i and j separated by a distance r_{ij}.

DISCUSSION

Fs pulse-train ablation of dielectrics

In first-principles calculation, an orthorhombic unit cell of SiO_2 is modeled with lattice constants of $9.29a.u.$, $16.09a.u.$ and $10.21a.u.$ along x, y and z axis respectively. Calculations are based on density functional theory (DFT) within the adiabatic local density approximation. The laser irradiation is modeled by subjecting the system to an external alternating electric field parallel to z axis. In addition, the ion position is frozen in all the calculations.

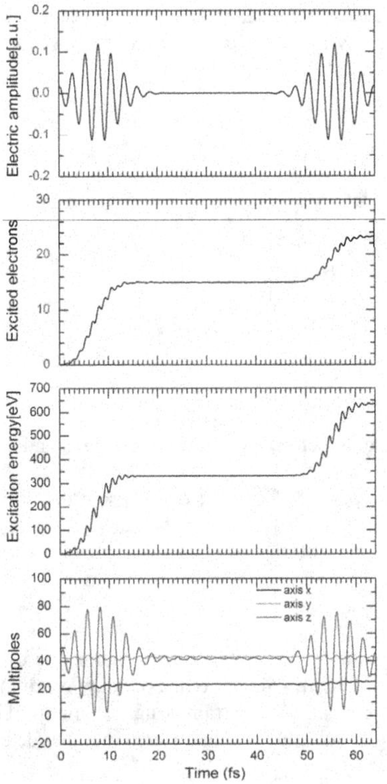

Figure 1. Laser electric field, excited electrons, excitation energy and multipoles.

Figure 1 shows the electron dynamics by fs pulse-train ablation of dielectrics, where the fs pulse-train consists of two sub-pulses with a separation time of 48 fs. Each of sub-pulse is a Gaussian wave packet with wavelength of 800nm, pulse duration of 16fs, peak power density of 5×10^{14} W/cm^2, as shown in figure 1a. Figure 1b shows that the number of excited electrons with time. At the initial stage, as laser electric field is weak, only a small amount of electrons are excited. However, an abrupt change occurs after 4fs, which is considered as a sign of the optical breakdown. Then the number of excited electrons reaches a first saturation value at around 12 fs during the first sub-pulse and a second saturation value at around 56 fs during the second sub-pulse. Figure 1c shows that the excitation energy or the total energy absorbed from laser electric field exhibits the similar evolution rules as the number of excited electrons. After laser irradiance, the total absorbed energy is 635.5eV and the average energy per excited electron is 27.38eV. Figures 1d shows that multipoles along z axis exhibit a 1-2 fs delay following laser electric field, whereas multipoles along x,y axis exhibit only slight disturbance. And multipoles reach the amplitude at around 8 fs and 56 fs with a smaller value of the second sub-pulse, which is due to the impact of a potential induced by free electron generations that weakens laser electric field.

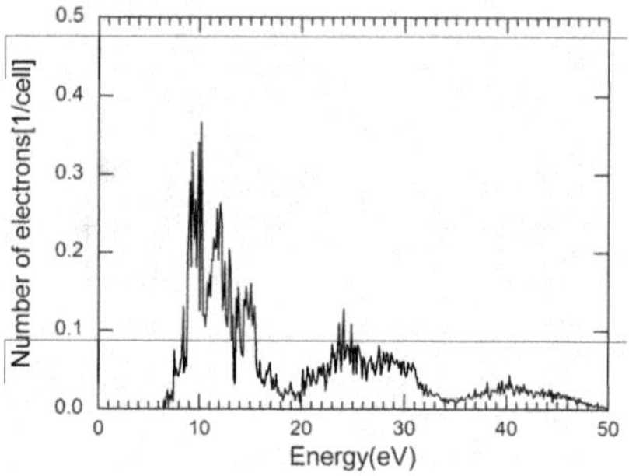

Figure 2. Occupation of excited electrons in the conduction band after laser irradiance.

Figure 2 shows the occupation of excited electrons in the conduction band after laser irradiance. As the laser power density (5×10^{14} W/cm^2 for each sub-pulse) is high enough for fully ionization of materials, the excited electrons are distributed broadly between 7-50eV.

Fs pulse-train ablation of metals

In the model combined two-temperature model and molecular dynamics simulation, nickel thin films are 18nm×2nm×200nm with periodic boundary conditions along x,y axis and free boundary conditions along z axis. Fs pulse-train design is employed by shaping single Gauss pulses to pulse-trains consisting of sub-pulses with different number/delay, as shown in figure 3.

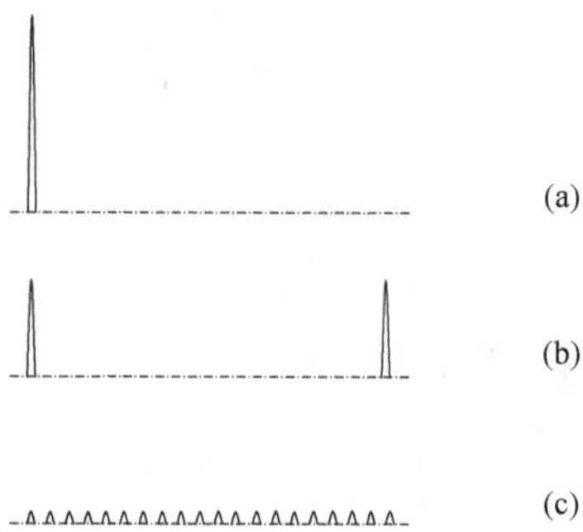

Figure 3. Schemes of fs pulse-train with the same total laser fluence. (a) classic fs pulse, (b) pulse-train with 2 sub-pulses with a separation time of 20 fs, (c) pulse-train with 20 sub-pulses with a separation time of 1 fs.

Figure 4 shows the snapshots of nickel thin films irradiated by fs pulse-train with the total laser fluence of 0.28J/cm^2, which indicates that fs pulse-train design dramatically changes electron dynamics and following energy transport and phase change processes [9]. For single pulse case, the nickel film fractures from inside accompanying several gas bubbles generation. For pulse-train with 2 sub-pulses case, the film fracture is weakened. For pulse-train with 20 sub-pulses case, the nickel film decomposes from the uppermost film without gas bubbles generation, which also leads to lower ablation rate with more and smaller uniform nanoparticles.

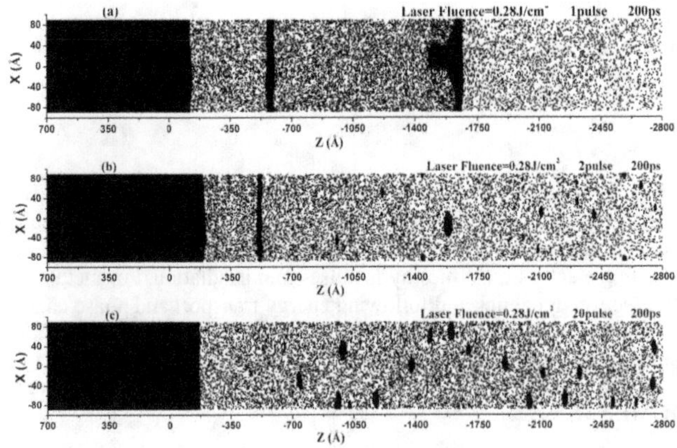

Figure 4. The snapshots of nickel thin films irradiated by fs pulse-train with the total laser fluence of 0.28J/cm^2. (a) classic fs pulse, (b) pulse-train with 2 sub-pulses with a separation time of 20 fs, (c) pulse-train with 20 sub-pulses with a separation time of 1 fs [9].

Figure 5 shows the time evolution of the dynamic film surfaces in ρ-T plane. For single pulse case, materials get across binodal line and approach spinodal line, where phase explosion is dominant mechanism. For pulse-train with 2 sub-pulses case, laser energy is inclined to deposit within film surface, and materials can get across both binodal and spinodal line. Still, phase explosion is dominant mechanism. For pulse-train with 20 sub-pulses case, materials can be heated to extremely high temperature and directly achieve supercritical state, where critical point phase separation is dominant mechanism.

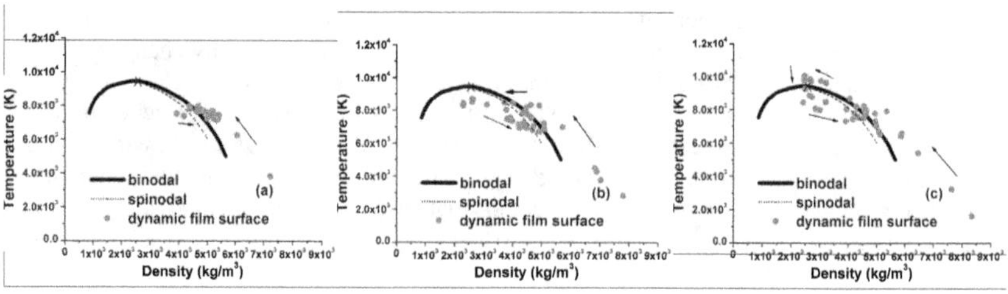

Figure 5. Time evolution of the dynamic film surfaces in ρ-T plane. (a) classic fs pulse, (b) pulse-train with 2 sub-pulses, (c) pulse-train with 20 sub-pulses [9]

CONCLUSIONS

This paper reviews our recent progress on the studies of changing electron dynamics to control the following properties and processes by fs pulse-train design for high-precision micro/nanoscale manufacturing. For dielectrics, first-principles calculation validates the feasibility of controlling electron dynamics by fs pulse-train irradiation. For metals, combined model predicts the electron dynamics and following energy transport and phase change processes can be dramatically changed by fs pulse-train design, which may favor high-precision micro/nanoscale manufacturing.

ACKNOWLEDGMENTS

This work was supported by the National Natural Science Foundation of China (NSFC) (Grants No. 90923039 and 51025521) and the 111 Project of China (Grant No. B08043).

REFERENCES

1. Y. Y. Cao, N. Takeyasu, T. Tanaka, X. M. Duan, and S. Kawata, *Small* 5, 1144 (2009).
2. A. Y. Vorobyev, and C. Guo,ion, *Phys. Rev. B* 72,195422 (2007).
3. Z. Hu, S. Singha, and R. J. Gordon, *Phys. Rev. B* 82, 115204 (2010).
4. L. Jiang, and H. L. Tsai, *Appl. Phys. Lett.* 87, 151104 (2005)
5. M. Spyridaki, E. Koudoumas, P. Tzanetakis, C. Fotakis, R. Stoian, A. Rosenfeld, and I. V. Hertel, *Appl. Phys. Lett.* 83, 1474 (2003).
6. R. Hergenröder, M. Miclea, and V. Hommes, *Nanotechnology* 17, 4065 (2006).
7. S. Noël and J. Hermann, *Appl. Phys. Lett.* 94, 053120 (2009).
8. L. Jiang, and H. L. Tsai, *Int. J. Heat Mass Transfer* 48, 487 (2005)
9. X. Li, L. Jiang, and H. L. Tsai, *J. Appl. Phys.* 106, 064906 (2009).
10. M. A. L. Marques, E. K. U. Gross, *Annu. Rev. Phys. Chem.* 55, 427 (2004).

Mater. Res. Soc. Symp. Proc. Vol. 1365 © 2011 Materials Research Society
DOI: 10.1557/opl.2011.1036

Nanomorphing with Ultrafast Lasers and Biomedical Applications

Alan J. Hunt
Department of Biomedical Engineering
University of Michigan,
Ann Arbor, MI, 48109 U.S.A.

ABSTRACT

Classically, the limit for optical machining is on the order of the wavelength of the incident light. However, by taking advantage of precise, nonlinear damage mechanisms that occur for femtosecond laser pulses, damage can be achieved on a scale an order of magnitude lower, allowing precise removal of very small amounts of material to produce holes mere tens of nanometers wide. Femtosecond laser nanomachining can be carried out in a variety of dielectrics, and in transparent substrates machining can be sub-surface, in contrast to other nanomachining techniques such as using an electron beam or focused ion beam. We focus on the use of glass, as it is in many ways an ideal material for use in biological applications due to its chemical, optical, electrical and mechanical properties. By precisely placing laser pulses in glass, three dimensional nano and microfluidic channels and devices can be formed including nozzles, mixers, and separation columns. Recent advances in this technique allow the formation of high aspect ratio nanochannels from single pulses, thus helping address the fabrication speed limitations presented by serial processing. These nanochannels have a range of applications including the fabrication of nanoscale pores and nanowells that may serve as vias between fluidic channels, or from channels to a surface. These nanochannels have applications as a standalone technique for fabrication of nanopores and nanowells, but can also complement other fabrication techniques by allowing precisely placed jumpers that can connect channels that are out of plane. We discuss applications for diagnostic microfluidic devices, and basic cell biology research.

INTRODUCTION

The macromolecular machinery that supports cellular functioning principally operates on the scale of nanometers, and nanoscale science and technology are broadly recognized as important emerging disciplines with broad potential to impact the development of new drugs and diagnostic devices. A principal challenge to realizing this potential is the difficulty of precise modification of materials at the micro and nanometer scales. This can be addressed by femtosecond laser nanomachining and ablation. Damage produced by femtosecond (fs) laser pulses is extraordinarily precise when the energy is near the damage threshold, that is at "critical intensity" (Joglekar et al., 2004). Indeed the precision of optics at critical intensity is so great that it enables laser machining of features that are far smaller than the limits that diffraction typically places on optical techniques (Du et al., 1996; Joglekar et al., 2003; Joglekar et al., 2004; Pronko et al., 1995; Temnov, 2006; Venkatakrishnan et al., 2002), and we have demonstrated precision down to the nanoscale (Joglekar et al., 2003; Joglekar et al., 2004), producing features as small as 20 nm in dielectrics. Nanometer scale features can also be fabricated in heterogeneous materials, including biological materials. Thus femtosecond nanomachining can produce features at a scale that competes with e-beam or ion-beam lithography, but is relatively simple to execute as it does not require a vacuum or special materials. Additionally, due to the sharp spatial confinement of femtosecond laser damage, chemical and structural changes are minimal beyond the ablated region. Furthermore, since this is an optical technique, it uniquely allows subsurface modification in optically transparent materials. By enabling such precise 3D fabrication or ablation, optical nanomachining constitutes a powerful tool for selectively ablating biological structures, and machining components for novel diagnostic and research devices.

EXPERIMENT

Here we present an overview of applications of femotosecond laser nanomachining and ablation that are being developed in our group. The following topics will be discussed:

1) Chemical separation. We have developed strategies to optimize debris clearing by expanding microbubbles produced during machining in liquids (Ke, 2005; Lee, 2007). This enables fabrication of long, thin (~0.5 micron) channels with aspect ratios greater than 1000:1. We have applied this to create an exceptionally small nanoscale capillary electrophoresis device that can perform a chemical separation using femtoliter sample volumes in under one second. Substances to be analyzed are introduced into the loading column by electroosmotic flow, and a fluid plug is pulled into the spiral separation column by switching the potential from

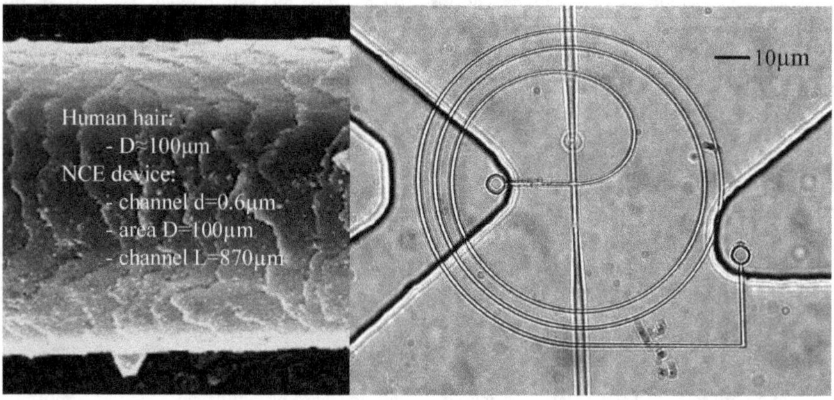

Figure 1. A functional nanoscale capillary electrophoresis device (right) compared with a hair (left). Vertical structure is a loading column, spiral is the separation column. Larger structures on right and left are the overlying PDMS header.

across the loading channel to across the separation column. As the analyte moves through the spiral, differential partitioning between fluid and solid phases (i.e. binding and unbinding to the channel walls) results in separation of substances as the fluid proceeds through the column.

2) Sensors and chemical analysis. The high-precision and repeatability of femtosecond laser damage allows machining of conical pores in glass, and these can be applied to detect

molecular aggregates by resistive-pulse sensing (An et al., 2008; Uram et al., 2006a; Uram et al., 2006b). Nanopores are immersed in an ionic solution, and a sample containing target molecules is placed on one side. When a potential is applied across a nanopore, the movement of molecular aggregates through the pore is detected as a resistive spike as current flow is impeded due to displacement of the conductive solution in the pore. Because the size of the spike increases with the size of the aggregate, target molecules can be detected by the increased size of molecular aggregates in the presence of a targeted crosslinker such as an antibody. The concentration can also be assessed since the frequency of spikes is proportional to the concentration of aggregates. We have applied this to directly detect proteins and viruses.

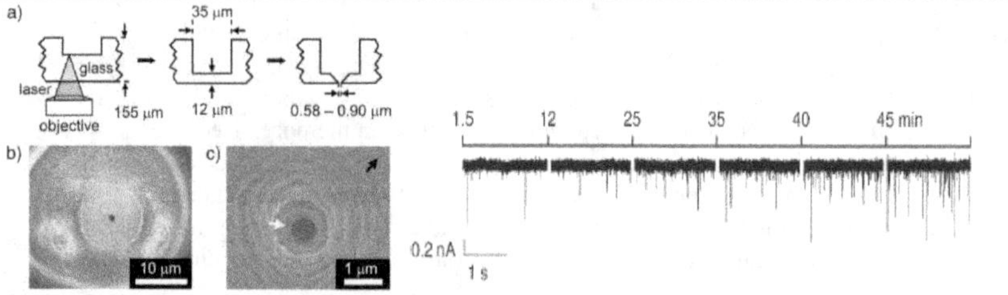

Figure 2. A nanopore in glass is used to detect biomolecule complexes by resistive-pulse sensing (Left), in this case a complex of antibodies and staphylococcal enterotoxin B (Right). Adapted from Uram et al (2006).

The ability to machine 3D structures in transparent substrates also enables direct inscription of components into existing microfludic devices. As a demonstration we machined an electrospray nozzle into the glass wall of a microfluidic chip (An et al., 2008).

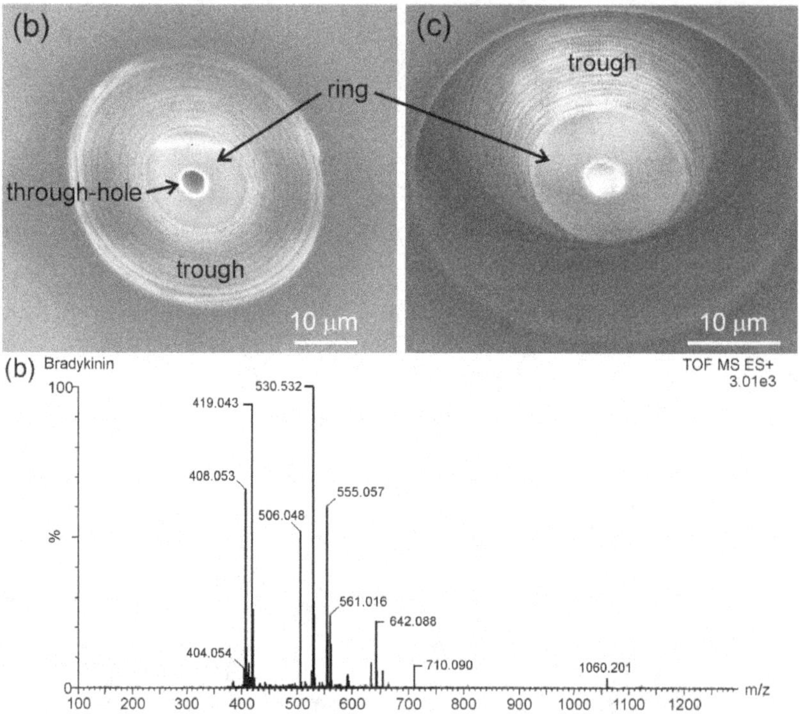

Figure 3. An electrospray nozzle is directly machined into a microfluidic device. Top: SEM images of a nozzle machined in glass. Bottom: Mass spectrum of bradykinin electrosprayed from the microfluidic device through the nozzle. Adapted from (An et al., 2008).

This allows the chip to directly interface with a mass spectrometer for analysis of the samples from within the chip.

3) Nanoscale liquid glass electrodes for actuating micro/nanofluidics. Actuating micro or nanoscale devices often requires delivery of electric currents to specific sites within a device. This is challenging due to the difficulty of introducing conductive wires through the structural substrate, especially if the conductors cannot be confined within a lithographic plane. We introduce a novel three-dimensional nanoscale electrode, produced by femtosecond laser nanomachining, and consisting of an electrolyte-filled nanochannel terminating at a glass tip a few hundred nanometers thick. We find that the thin glass wall transforms from an insulator to conductor when the potential exceeds a modest threshold potential of a few volts. Remarkably this process is reversible; we argue that this transition is an intrinsically nanoscale phenomenon, which depends on the high resistance of the nanochannels to stabilize currents, and extremely fast thermal dissipation at the nanoscale tip to prevent thermal damage during dielectric breakdown. We proposed that conductivity occurs through dielectric breakdown, analogous to a Zener diode (Lee et al., 2010). Alternatively, a similar phenomenon, identified though a different approach, has been attributed to glass behaving as a hydrogel for nanoscale thickness (Velmurugan et al., 2010).

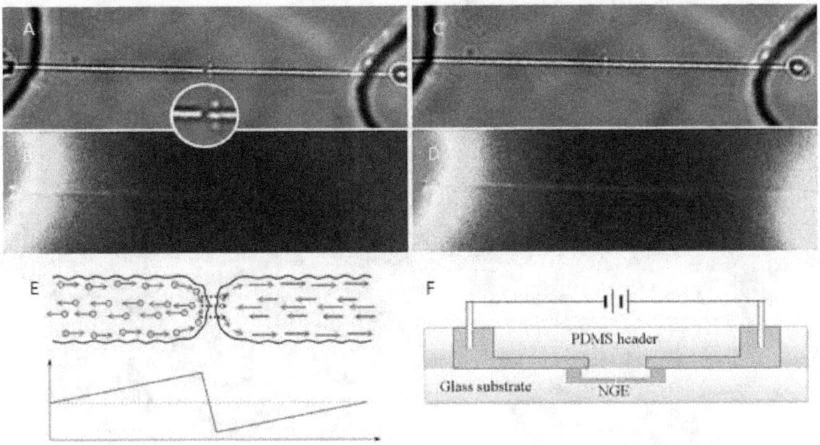

Figure 4. Electric current measurements and fluid flow visualization across a nanometer scale liquid-glass electrode (NLGE). (a) Light micrograph of the NLGE carrying a current of 18.4 nA with 10 V potential; (b) Despite current flow, the nanoglass wall blocks fluid flow, as evidenced by lack of rhodamine 110 (R110, 40 μM) fluorescence on the right side (pseudo-color conversion of grey-scale to black-blue-green-yellow-red format); (c) elimination of the nano-glass wall increases the electric current to 53.4 nA, and (d) fluid flows to the right reservoir; (e) schematics of flow fields and resultant electroosmotic pressure generation; (f) schematic diagram of the microchip for testing the NLGE.

We apply reversible nanoscale glass conductivity to fabricate an extraordinarily small nanoinjector, which is driven by an electrokinetic pump actuated by a nanoscale liquid glass electrode.

4) Single-shot machining of high-aspect nanochannels. A significant impediment to widespread application of femtosecond laser machining is relatively slow fabrication speeds associated with serial processing. Recent work by our group and others has demonstrated that high aspect ratio nanochannels can be fabricated with single femtosecond laser pulses (Herbstman & Hunt, 2010; Kudryashov et al., 2007; White et al., 2008). These channels, which we argue are formed by microfilimentation during the breakdown process, enable rapid production of long channels for nanopores or fluidic connections. With advances in laser technology allowing high repetition rate, high energy femtosecond lasers (Keller, 2003), formation of long channels using single pulses could allow fabrication speeds as high as 400 m/s. Thus femtosecond laser machining rates would approach rates competitive with photolithographic techniques, while enabling nanoscale and three dimensional features to be formed outside of

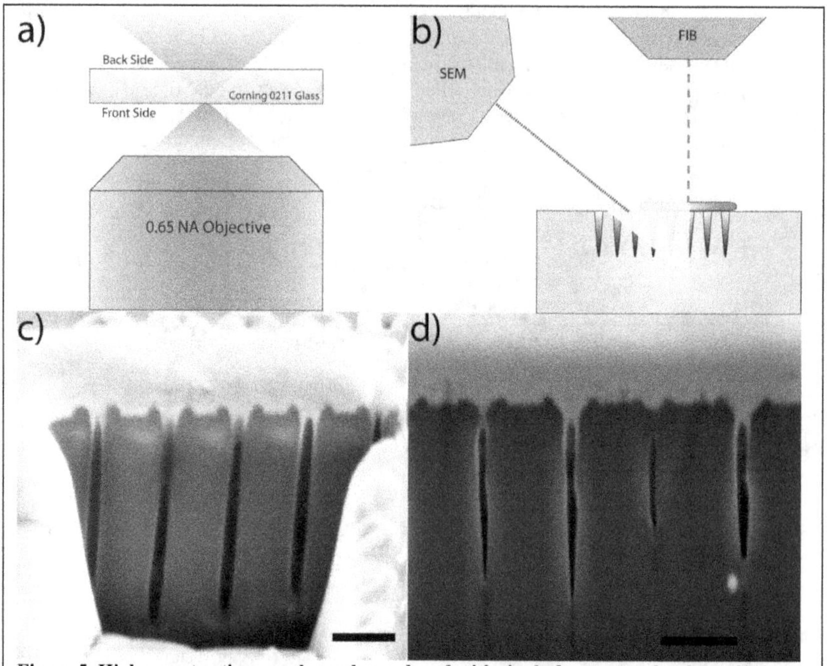

Figure 5. High aspect ratio nanochannels produced with single femtosecond pulses. A) An illustration of the machining geometry. B) A damaged sample examined under a dual beam FIB/SEM. The FIB is placed perpendicular to the sample surface and the SEM is offset at a 52° angle. The sample is covered with a few hundred nanometers of platinum to protect the surface from unintended ion damage. The FIB removes a wedge from the sample, exposing the nanochannel cross section. C) An example cross section for back side machining. D) An example cross section for front side machining. Both images show damage at 67 J/cm^2. Scale bars indicate 1 μm.

a clean room. This phenomenon also holds promise for industrial applications ranging from large arrays of nanowells to both in-plane and out-of-plane modification of patterned structures, performed entirely with a single optical machining device (McDonald, 2006).

The high aspect ratio also relaxes the requirements for nanoscale calibration between the laser focus and a surface when holes are drilled, providing a useful way to fabricate the initial few microns of the pores that interface with the surface and a subsurface channel. This relaxes the tolerances for machining the subsurface channel, which only need come within an easily achieved few microns of the surface to connect with the pores. The reproducibility is excellent, and the mouth of the channels forms a smooth, slightly raised lip reminiscent of glass patch pipettes and ideal for forming patch-clamp gigaseals, suggesting this may be a useful approach for constructring glass chips for probing electrical and chemical properties of cells on a substrate suitable for high quality microscopy.

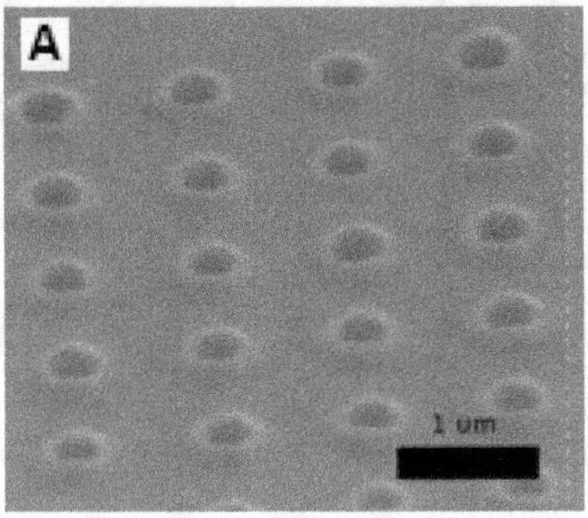

Figure 6. SEM of an array of high-aspect-ratio nanoholes formed in glass by single laser pulses. An array of holes formed by translating the stage between laser shots.

DISCUSSION

Femtosecond laser micro and nanomachining provides a flexible tool for 3D machining in transparent dielectrics. Highly precise surface and subsurface machining enables direct fabrication of high aspect ratio channels, nozzles, and nanopores for biomedical applications including biochemical separation and analysis for diagnostics and biosensors. 3D fs-laser machining creates at least two intriguing opportunities for complex flow control. The first is that the extremely narrow-bore channels make possible much more complex valveless, electrokinetically-driven microfluidics, for example the long-sought ability to individually-address (i.e. inject or remove fluid from) arrays of reaction chambers: multiplexing more than two external sources to inject fluids into each microwell in a 2D array is topologically impossible in a plane without crossing streams, since at least two inlets and one outlet is needed.

Because of the current limits on processing speed imposed by serial production methods such as laser machining, initially we envision that this approach would be used as an adjunct to wet-etching and lithographic approaches. In this manner, traditional wet-etching produces the majority of planar features needed in a microchip, followed by laser-machining to create 3D features such as jumpers, mixers, and other out-of-plane features. An analogous process is used in the integrated circuit industry, which routinely uses laser-trimming to trim resistors and other alterations after the bulk of a chip is produced using photolithographic fabrication processes. With continued advances in high average laser power lasers and laser machining methodologies, femtosecond laser machining will advance to meet the requirements for complete fabrication of

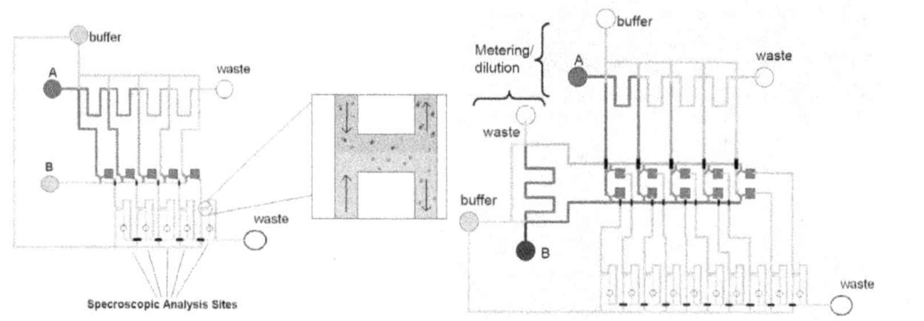

Figure 7. Example application: an integrated immunoassay device. Microscale dilution array and diffusion analysis system are integrated to make a self-contained lab on a chip that prepares, mixes, and analyses antigen/antibody binding. Upper diagram: in the simplest design fluorescently-labeled antibody solution A is serially diluted and mixed with a single dilution of the analyte (solution B) for analysis of antigen content. Antibody-target conjugate are separated from free antibodies by diffusion across a laminar flow in an "H" channel, and free antibody concentrations are assessed by spectroscopic analysis. Lower diagram: A somewhat more complex design assays two concentrations of analyte.

increasingly complex forms, starting from monolithic substrates and outside of the cleanroom.

ACKNOWLEDGMENTS

This work was supported by a grant from IMRA Corp., NIH R21 EB006098, and NIH R01 GM072006.

REFERENCES

An, R., Hoffman, MD, Donoghue, MA, Hunt, AJ, Jacobson, SC. 2008. Water-assisted femtosecond laser machining of electrospray nozzles on glass microfluidic devices. *Optics Express*. 16:6.

An, R., J.D. Uram, E.C. Yusko, K. Ke, M. Mayer, and A.J. Hunt. 2008. Ultrafast laser fabrication of submicrometer pores in borosilicate glass. *Opt Lett*. 33:1153-5.

Du, D., X. Liu, and G. Mourou. 1996. Reduction of multi-photon ionization in dielectrics due to collisions. *Appl. Phys. B*. 63:617-621.

Herbstman, JF, and AJ Hunt (2010) High-aspect ratio nanochannel formation by single femtosecond laser pulses. Optics Express, 18: 16840-16848.

Joglekar, A.P., H. Liu, G.J. Spooner, E. Meyhofer, G. Mourou, and A.J. Hunt. 2003. A study of the deterministic character of optical damage by femtosecond laser pulses and applications to nanomachining. *Applied Physics B-Lasers and Optics*. 77:25-30.

Joglekar, A.P., H.H. Liu, E. Meyhofer, G. Mourou, and A.J. Hunt. 2004. Optics at critical intensity: Applications to nanomorphing. *Proceedings of the National Academy of Sciences of the United States of America*. 101:5856-5861.

Ke, K., Hasselbrink, E., Hunt, A.J. . 2005. Rapidly-Prototyped Three-Dimensional Nanofluidic Channel Networks in Glass Substrates. *Anal. Chem.* 77:5.

Keller, U. 2003. Recent developments in compact ultrafast lasers. *Nature*. 424:831-8.

Kudryashov, S., Mourou, G, Joglekar, A, Herbstman, JF, Hunt, AJ. 2007. Nanochannels fabricated by high-intensity femtosecond laser pulses on dielectric surfaces. *Applied Physics Letters*. 91.

Lee, S., Bull, JL, Hunt AJ. 2007. Acoustic phenomena during femtosecond laser machining. *Applied Physics Letters*. 91.

Lee, S., An, R, Hunt AJ, 2010. Liquid Glass Electrodes for Nanofluidics. *Nature Nanotechnology*, 5: 412 – 416.

McDonald, J., Mistry, VR, Ray, KE, SM Yalisove. 2006. Femtosecond pulsed laser direct write production of nano- and microfluidic channels. *Applied Physics Letters* 88.

Pronko, P., S. Dutta, J. Squier, J. Rudd, D. Du, and G. Mourou. 1995. Machining of Sub-Micron Holes using a Femtosecond Laser at 800 nm. *Opt. Commun.* 114:106-110.

Temnov, V., Sokolowski-Tinten, K, Zhou, P, El-Khamhawy, A, von der Linde, D 2006. Multiphoton Ionization in Dielectrics: Comparison of Circular and Linear Polarization. *Phys. Rev. Lett.* . 97:4.

Uram, J.D., K. Ke, A.J. Hunt, and M. Mayer. 2006a. Label-free affinity assays by rapid detection of immune complexes in submicrometer pores. *Angew Chem Int Ed Engl*. 45:2281-5.

Uram, J.D., K. Ke, A.J. Hunt, and M. Mayer. 2006b. Submicrometer pore-based characterization and quantification of antibody-virus interactions. *Small*. 2:967-72.

Velmurugan, J., D. Zhan, and M.V. Mirkin. 2010. Electrochemistry through glass. *Nature Chemstry* 2:498

Venkatakrishnan, K., B. Tran, P. Stanley, and N. Sivakumar. 2002. The effect of polarization on ultrashort pulsed laser ablation of thin metal films. *J. Appl. Phys.* 92:1604-1607.

White, Y., Li, XX, Sikorski, Z, Davis, LM, Hofmeister, W. 2008. Single-pulse ultrafast-laser machining of high aspect nano-holes at the surface of SiO2. *Optics Express* 16:10.

Mater. Res. Soc. Symp. Proc. Vol. 1365 © 2011 Materials Research Society
DOI: 10.1557/opl.2011.1221

Laser direct writing of nanocompounds

Andreas Ostendorf [1], M'Barek Chakif [1], Qingchuan Guo [1]

[1] Ruhr-University Bochum, Laser Applications Technology, Universitaetsstr. 150, D-44801 Bochum, Germany

ABSTRACT

Laser direct polymerization has been proven as a powerful tool to generate microstructures. Often photosensitive polymer materials are used because they can be tuned by photoactive molecules to be susceptible to a specific wavelength of light to initiate the polymerization process. One of the main drawbacks of this technique is the lack of functional polymers, e.g. conductive, magnetic, mechanical, optical or bioactive materials. Nanocomposites (nanocompounds), i.e. polymers with inorganic nanomaterials incorporated in the matrix offer a huge variety of new functionalities. A new approach will be presented how functional nanocomposite polymers can be generated and used for laser direct writing techniques. This can open the door for completely new MEMS and MOEMS devices comprising active and passive subcomponents.

INTRODUCTION

A nanocomposite is defined as a material of more than one solid phase, where at least one dimension falls into the nanometer scale [1]. The fabrication of nanocomposites opens up an attractive route to obtain novel, optimized, and miniaturized compounds that can meet a broad range of applications.

Nowadays polymers play a very important role in numerous fields of everyday life due to their advantages over conventional materials (e.g. wood, clay, metals) such as lightness, resistance to corrosion, and low cost production. Besides, photosensitive polymers provide good processability that can be cured at a specific light wavelength. Another advantage is that polymers are easy to handle and have many degrees of freedom for controlling their properties. Further improvement of their performance, including composite fabrication, still remains under intensive investigation. The altering and enhancement of the polymer's properties can be achieved by doping with various nanoparticles such as metals, semiconductors, magnetic, organic and inorganic particles and fibres, as well as carbon structures and ceramics [2-6]. Such additives are used in polymers for a variety of purposes, for example: improved processing, density control, optical effects, thermal conductivity, control of the thermal expansion, electrical properties, magnetic properties, flame resistance, and improved mechanical properties, such as hardness, elasticity, and tear resistance [7-10]. Polymers with various functional properties can be synthesized using additives. Laser direct polymerization [11, 12] (e.g. single-, two-photon polymerization) has been proven as a powerful technique for the fabrication of micro- and nanostructures. The functional polymers will be used to write structures by laser direct polymerization. This technology opens new prospects for the realization of novel MOEMS with increased functionality, integration, and high level of miniaturization.

We present our investigations into the fabrication of functional structures by laser direct polymerization on SWNTs (Single-Walled Carbon Nanotubes) and TiO_2 -Polymer nanocomposite. For this purpose we pursue two different strategies. The first strategy is to produce and investigate nanoparticle-based nanocomposites as candidates for synthesis of functional composites, and the second strategy is via laser direct writing technique to write functional structures.

EXPERIMENTAL DETAILS

High conductivity of SWNTs (Nanoamor Inc. $\sigma > 10^4$ S/m) have been doped into the low conductivity polymer (Ormocore® with 1.8% Photoinitiator Irgacure from Microresist technology GmbH $\sigma \approx 10^{-14}$ S/m) matrix to tailor the electrical properties suitable for different applications such as microwires, because different applications normally require specific levels of conductivity. Very high refractive index (n=2.6-2.9, depending on the crystal axis [13]) of TiO_2 (Evonik GmbH) into the polymer ($n_{polymer} = 1.554$) matrix can improve the composite's refractive index for the fabrication of microoptical components, micronanomachines, and their integrated systems by means of 2PP. The high refractive index of composite will be used e.g. in optical tweezers, as the acting forces depend on the refractive index of the material [14].

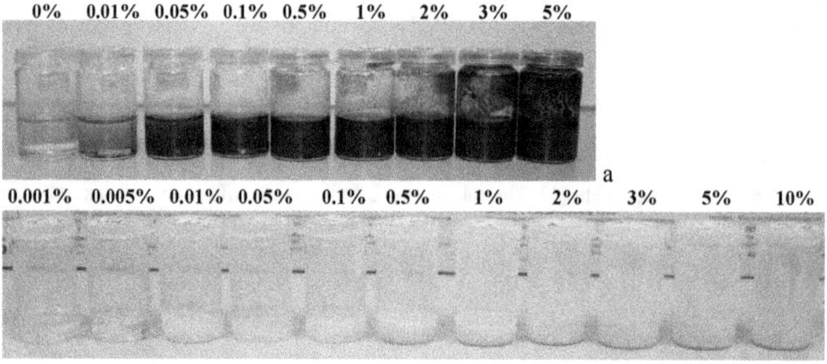

Figure 1. Photograph of SWNTs/polymer composite (a) and TiO_2/polymer composite (b) at different concentration.

SWNTs/polymer composites with a concentration range between 0.01 and 5.0 wt.-%, and TiO_2/polymer composites with a concentration from 0.001-10 wt.-% were prepared (Figure 1) by adding pure nanoparticles to polymer (ex-situ preparation). The mixtures were first stirred for 24h at 70°C by magnetic stirring, and then sonicated by a bath-type sonicator for different duration at 50°C. SWNTs/polymer composites were sonicated for 60min, 75min, 90min, 105min and 300min, and TiO_2/polymer composites with sonication time of 100min and 160min were prepared. Subsequently, the mixtures were sonicated again using a bath-type sonicator for 15min at 20°C room temperature. A small drop of the mixture was placed between two polyester films where the distance (200μm) between the two films was controlled by a spacer film placed between them. UV light from an extra high-pressure mercury lamp immediately irradiated the mixture for various times (1-10min) depending on the concentration of the nanoparticles.

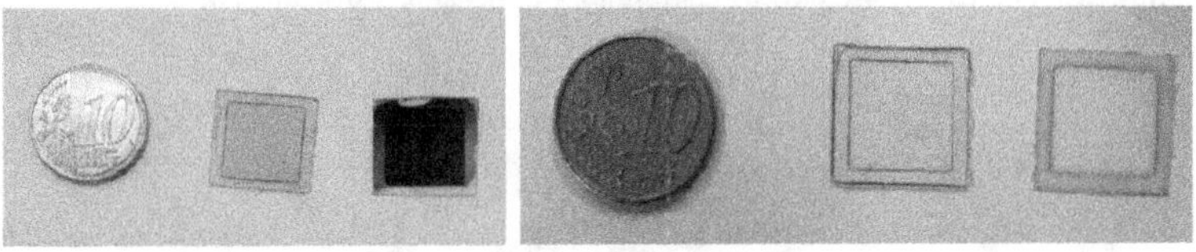

Figure 2. Photograph of SWNTs-composite (left) and TiO2-composite (right) film at low (0.01%) and high (3%) concentration. The 10 Eurocent coin is show to compare the sample size.

The composite films (Figure 2) were characterized after peeling them off from the polyester films. UV-Vis-NIR and Raman spectroscopy measurements were recorded using JASCO spectrophotometer and Renishaw Raman Microscope, respectively. Refractive index measurements were carried out by a refractometer (Carl Zeiss AG), and four-point probe was used to measure the electrical conductivity of the composites. After synthesizing of the nanocomposites with different concentrations, 1- and 2PP processing were used to write the functional structures.

RESULTS AND DISCUSSION

UV-Vis-NIR spectroscopy measurements

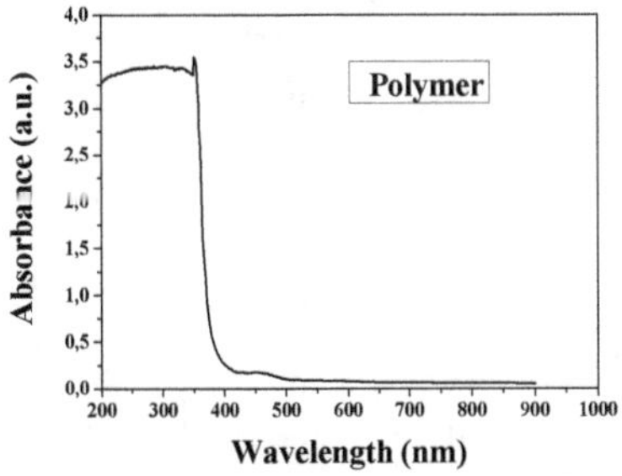

Figure 3. Absorption spectrum of pristine polymer (Ormocore® incl. 1.8% Photoinitiator). At the wavelength λ =355nm high absorption can be observed.

Figure 3 shows the absorption spectrum of the solidified film from pristine polymer of thickness 200µm. It is obvious that at a wavelength of 355nm (3ω of common DPSS lasers) the polymer exhibits a high absorption whereas the absorption rate decreases with increased wavelength. Using UV-Vis-NIR measurement to search for the optimum wavelength for 1PP in order to allow 2PP processing, moreover, when the laser beam can transmit through the composite, at 2PP wavelength the composite muss show high transmission. Only under these circumstances the polymerization area can be three-dimensionally localized at the focal point of laser [15, 16]. In this case 2PP can be used to write arbitrary 3-dimensional structures in composites.

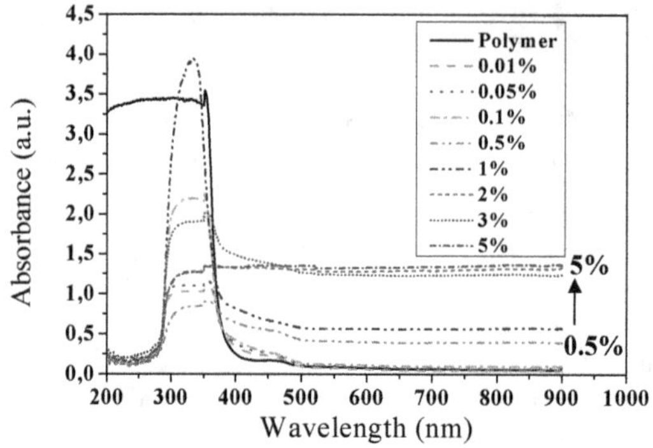

Figure 4. Absorption spectra of SWNTs/polymer composites after sonicatiom time 105min with different concentrations. All samples show high absorption at 355nm, whereas the absorption in the Vis-NIR increased for concentrations higher than 0.5 wt.-%.

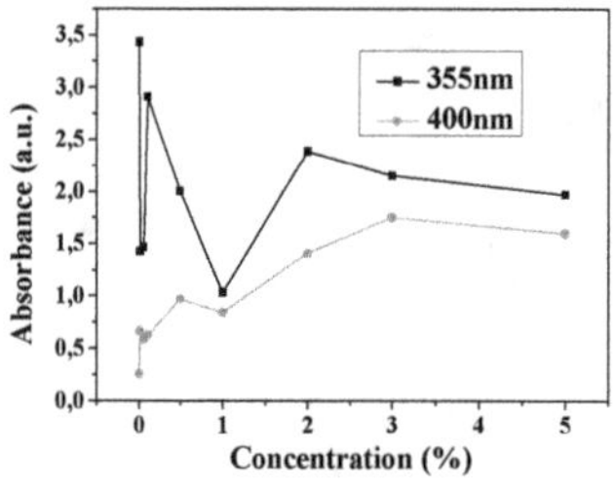

a

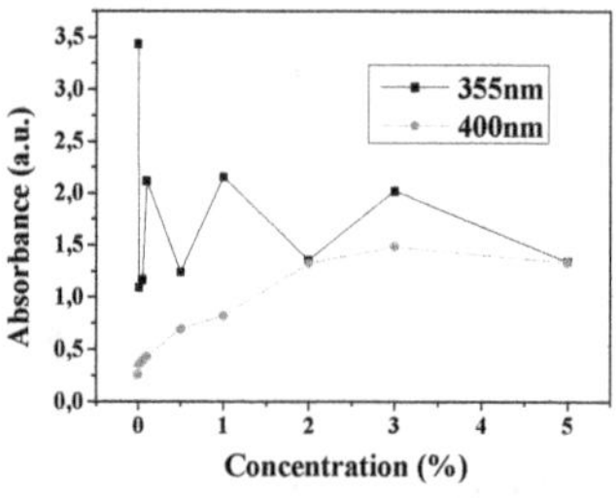

b

Figure 5. Absorption spectra of SWNTs/polymer composites at 355nm and 400nm with different sonication time (a: 60min, b: 105min).

UV-Vis-NIR spectroscopy measurements for the polymerized samples are depicted in Figure 4. It is clearly shown that at 355nm all samples exhibit extensive absorption although from 0.5 wt.-% the absorption also increases in the Vis-NIR area. The absorbance at 355nm and 400nm were compared for all the samples, as shown in Figure 5. It can be seen that all the samples have higher absorption at 355nm compared to 400nm which is the 2ω wavelength of common TiSa-lasers. Therefore, in future work 710nm will be chosen as the optimum wavelength for 2PP processing.

Refractive index measurements

The refractive index of the TiO_2-nanocomposite can be modified, as shown in Figure 6. It can be seen that the refractive index of the composite roughly follows a linear function of the TiO_2 concentration in the composite, indicating that the doping concentration is an important factor. However, in high concentration range above 0.5 wt.-%, the transmission rate of the composite decreases. Consequently the 2PP wavelength cannot transmit through the composite anymore which makes it difficult to write full 3D structures. Thus, a suitable composite with higher refractive index and higher transmission has to be developed.

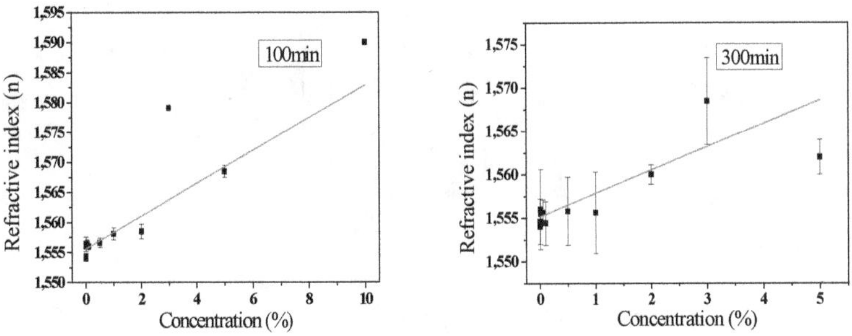

Figure 6. Refractive index of the TiO_2/polymer composites with different sonication times from 100min (left) to 300min (right) as a function of TiO_2 concnetration at a wavelength of 589.6nm.

Raman spectroscopy measurements

In order to characterize the influence of the polymer matrix on the unique properties of the SWNT, especially as the electrical properties of the SWNT after mixing and polymerization might be changed, Raman spectra of the two components were recorded in advance and after processing (Figure 7). The Raman spectra of the processed samples at an excitation wavelength of 785nm are shown in Figure 8 and 9. The spectra are almost identical to the superposition of pristine polymer and raw SWNTs. The top right figures in Figure 8 and 9 for the G-Band area show that all samples exhibit almost the same peak. The result shows that grafting of the polymer onto the sidewall of the SWNTs did not occur and the SWNTs remain intact during the polymerization in the present system.

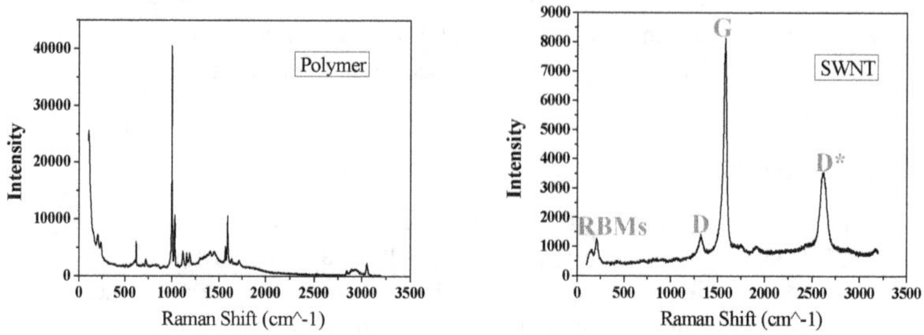

Figure 7. Raman spectra of pristine polymer (left) and raw SWNTs (right): The characteristic peaks correspond to diameter dependent radial breathing mode (RBM) and disorder-induced D and tangential G band of SWNTs.

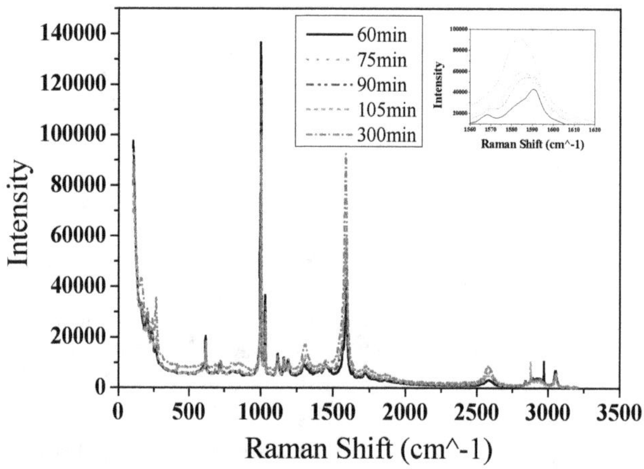

Figure 8. Raman spectra of SWNTs (1 wt.-%)-composite at different sonication time. The top right figure shows in G-Band area all samples at different sonication time have almost the same peak.

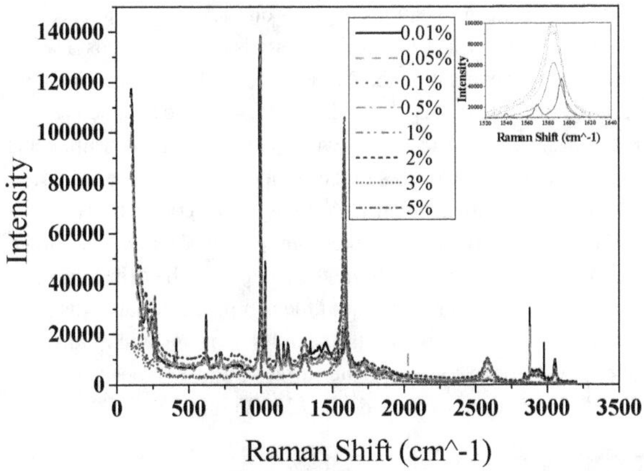

Figure 9. Raman spectra of SWNTs/polymer composites at different concentration. The top right figure shows in G-Band area all samples with different concentration have almost the same peak.

Conductivity measurements

C_i/wt% t/min	0.01	0.05	0.1	0.5	1	2	3	5
60min	X	X	X	X	X	X	X	X
75min	X	X	X	X	X	X	X	X
90min	X	X	X	X	X	X	X	X
105min	X	X	X	X	X	X	X	1,29E-07 S/m
105*min	X	X	X	X	X	3,81E-08 S/m	2,53E-07 S/m	9,70E-06 S/m
300min	X	X	X	X	X	3,57E-08 S/m	9,05E-07 S/m	3,03E-06 S/m

Table 1. Conductivity measurements of all samples with different sonication time and concentration: after polishing the surface of composite films at the sonication time 105*min, at high concentration the conductivity increased. X: beyond measuring range of device.

The effective use of SWNTs in composite applications depends on the ability to disperse the SWNTs homogeneously throughout the matrix. Electrical conductivity is one of the most useful properties of SWNT/polymer composites for evaluating the degree of the SWNT dispersion and for developing possible applications of the composites [17, 18]. In this paper the conductive SWNTs/polymer composite was synthesized to fabricate nanowires. Conductivity measurements can be used to examine the dispersion of the nanoparticles in the composite. Therefore we fabricated several nanowires for applications with different specific levels of conductivity. Four-point probe was used to measure the conductivity of all samples after polymerization. As shown in Table 1, only at high concentration and long dispersion time the composites have become conductive. The conductivity was found higher when the film of the composite surface was polished with a removal layer of about 10μm, as shown for the samples at a sonication time of 105*min. This indicates that the for polymerized composite the surface of the composite is purely polymer whereas most of SWNTs are dispersed inside the film (Figure 10).

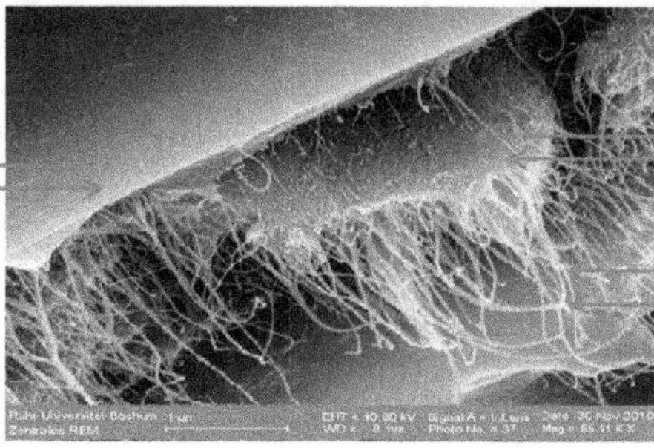

Figure 10. Cross-section of polymerized SWNTs-composite after breaking, most of SWNTs are localized inside the film.

Single- and Two-Photon Polymerization Results

Single-Photon Polymerization (1PP) Results

Figure 11 shows the principle setup for the fabrication of 2-dimensional structures by 1PP. The pristine polymer and all samples were written by means of 1PP using a picosecond UV laser (LUMERA LASER GmbH; wavelength $\lambda = 355$nm, repetition rate f = 200KHz). The laser fluence was 70.8 mJ/cm^2, writing speed was set to 10mm/s.

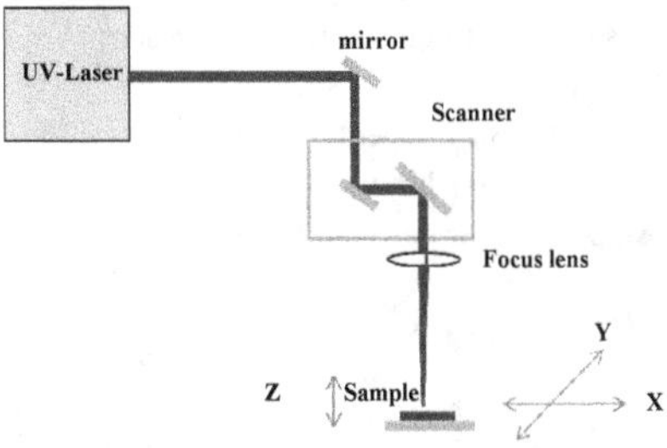

Figure 11. Principle setup for the fabrication of 2D structures by 1PP.

Composites within the investigated concentration range could be properly polymerized. Figure 12 and 13 show the structures at low and high concentration of composites. The surface of the structures from composites at low concentration is perfectly smooth. In contrast, composites with higher concentration show a rough surface indicating that there are some nanoparticles agglomerates on the surface of structures. In this case the laser beam was heavily scattered by nanoparticles which can damp the polymerization process, especially for 3-dimensional processing where transmission at the fundamental laser wavelength is required.

Figure 12. SEM images of SWNTs-composite structures at 0.01 wt.-% (a) and 3 wt.-% (b).

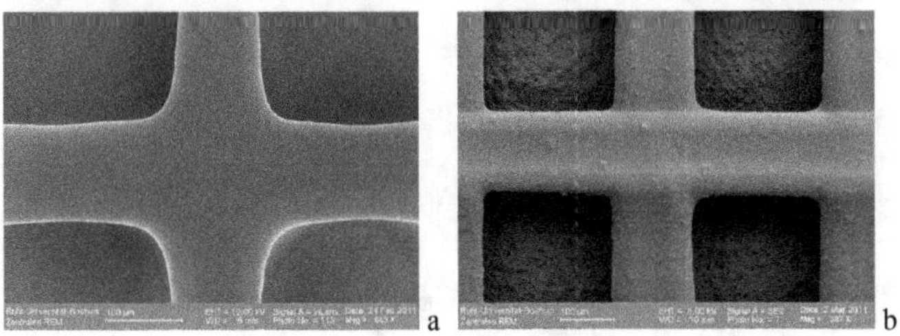

Figure 13. SEM images of TiO_2-composite structures at 0.001 wt.-% (a) and 2 wt.-% (b).

Two-Photon Polymerization (2PP) Results

The experimental setup used for 2PP photofabrication is shown in Figure 14. In the present work, for processing SWNTs/polymer composite, a femtosecond laser system (Spectra Physics Tsunami®, wavelength λ = 800nm, delivering pulses of 90-fs duration at repetition rate f = 80MHz) was used. A 100× microscope objective lens (Nikon, Plan Apochromat, N.A. = 1.4) was used to focus the laser beam into the volume of the composite. The laser power was set to 32.6mW, writing speed was 4mm/s.

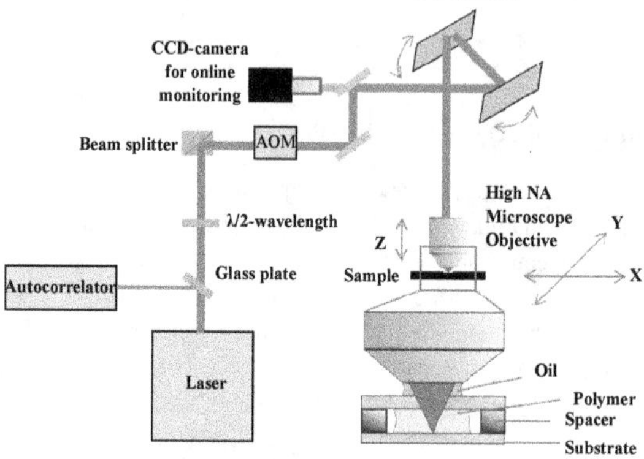

Figure 14. Experimental setup used for 2PP photofabrication.

Figure 15 shows SEM image of SWNTs/polymer composite structures fabricated by means of 2PP. Some structures have been shrunken, which have led to the distortion of the fabricated structures. In order to avoid such distortion, in future work more suitable laser parameters will be chosen, in addition the polymerization process should be initiated at focus point on the interface.

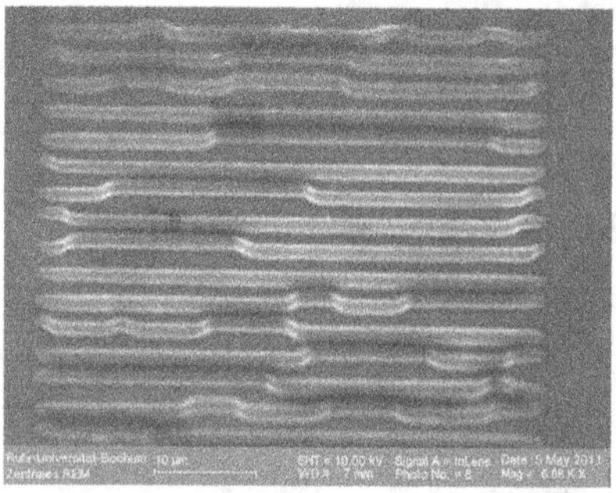

Figure 15. SEM image of SWNTs/polymer composite structures at concentration 0.01 wt.-%.

CONCLUSIONS

We have presented investigations into functional composites which can be structured by laser direct writing. The TiO_2/polymer composites, by varying its TiO_2-nanoparticles concentration, can be used to "tune" its refractive index. For SWNTs/polymer composites, by increasing the SWNTs concentration and sonication time, the conductivity can principally be adjusted. UV-Vis-NIR measurements show that at the common UV laser wavelength of 355nm all samples show high absorption. The optimum wavelength for 1PP processing is at 355nm, for further 2PP processing the optimum wavelength is set to 710nm for the future work. However, the transmission of all composites decreased in the Vis-NIR spectrum for concentrations higher than 0.5 wt.-%. In addition Raman spectroscopy have shown that the spectra of SWNTs/polymer composites are almost identical to the superposition of pristine polymer and raw SWNTs. The composites were photopolymerized both at low and high concentration by 1PP processes. At high concentration the laser beam was scattered by nanoparticles. It can be concluded that extinction of embedded nanoparticles by absorption and scattering will damp the polymerization process, especially for 3-dimensional structures based on 2PP processing. Using 2PP processing we have produced composite structures, however, the structures were still distorted.

In future, we will apply some of the presented methods to investigate and determine the boundaries of nanoparticles concentration and 2PP processing. In this way we will gain the suitable nanoparticles concentration that the composites can be used both as functional and structural polymers.

ACKNOWLEDGMENTS

We would like to thank Deutsche Forschungsgemeinschaft (DFG) for the generous support of this work in the framework of projects C11 (SFB 459) and OS188/17-2.

REFERENCES

1. L. E. Nielsen, R. F. Landel, "Mechanical Properties of Polymer and Composites," 2nd ed.; *Marcel Dekker*: New York, NY, USA, (1994).
2. S. Glushanin, V. Y. Topolov, A. V. Krivoruchko, "Features of piezoelectric properties of 0-3 $PbTiO_3$-type ceramic/polymer composites," *Materials Chemistry and Physics* **97 (2-3)**, 337-364, (2006).
3. P. Hine, V. Broome, I. Ward, "The incorporation of carbon nanofibres to enhance the properties of self reinforced, single polymer composites," *Polymer,* **46 (24)**, 10936-10944, (2005).
4. N. Cioffi, L. Torsi, N. Ditaranto. G. Tantillo, L. Ghibelli, T. Sabbatini, T. Bleve-Zacheo, M. D`Alessio, P. G. Zambonin, E. Traversa, "Copper nanoparticle/Polymer composites with antifungal and bacteriostatic properties," *Chemistry of Materials* **17 (21)**, 5255-5262, (2005).

5. A. Pelaiz-Barranco, P. Marin-Franch, ''Piezo-, pyro-, ferro-, and dielectric properties of ceramic/polymer composites obtained from two modifications of lead titanate,'' *Journal of Applied Physics* **97 (3)**, 034104-034104-5, (2005).
6. Juan Wang, Hong Xia, Bin-Bin Xu, Li-Gang Niu, Dong Wu, Qi-Dai Chen, and Hong-Bo Sun, ''Remote manipulation of micronanomachines containing nanoparticles '', *Opti. Letter.* **34**, 5 (2009).
7. M. Meyyappan, ''Carbon nanotubes science and applications,'' CRC Press: *Boca Raton*, FL, (2005).
8. Z. M. Huang, Y. Z. Zhang, M. Kotaki, S. Ramakrishna, ''A review on polymer nanofibers by electrospinning and their applications in nanocomposites,'' *Composites Science and Technology* **63 (15),** 2223-2253, (2003).
9. J. Jordan, K. I. Jacob, R. Tannenbaum, M. A. Sharaf, I. Jasiuk, ''Experimental trends in polymer nanocomposites – a review,'' *Materials Science and Engineering A-Structural Material Properties Microstructure and Processing* **393 (1-2),** 1-11, (2005).
10. J. F. Gerard, ''Fillers and filled polymers,'' Wiley-VCH: *Weinheim*, (2001).
11. H. B. Sun and S. Kawata, Adv. *Polym. Sci.* **170**, 169 (2004).
12. S. Maruo, O. Nakamura, and S. Kawata, *Opt. Lett.* **22**, 132 (1997).
13. D. R. Lide, ''Handbook of Chemistry and Physics,'' 76th edition, CRC Press, *Boca Raton*, Florida (1995).
14. A. Ashkin, JM. Dziedzic, JE. Bjorkholm, S. Chu, ''Observation of a single-beam gradient force optical trap for dielectric particles,'' *Opt. Lett.* **11**: 288-290 (1986).
15. S. Kawata, H: B. Sun, T. Tanaka, and K. Takada, ''Finer features for functional microdevices,'' *Nature* **412**, 697-698 (2001).
16. S. Juodkazis, V. Mizeikis, K. K. Seet, M. Mima, and H. Misawa, ''Two-photon lithography of nanorods in SU-8 photoresist, '' *Nanotechnology* **16**, 846-849 (2005).
17. Y. J. Kim, T. S. Shin, H. D. Choi, J. H. Kwon, Y. C. Chung, H. G. Yoon, ''Electrical conductivity of chemically modified multiwalled carbon nanotube/epoxy composites,'' *Carbon 43 (1)*, 23-30, (2005).
18. S. Barrau, P. Demont, A. Peigney, C. Laurent, C. Lacabanne, ''DC and AC conductivity of carbon nanotubes-polyepoxy,'' *Macromolecules* **36 (14)**, 5187-5194, (2003).

Mater. Res. Soc. Symp. Proc. Vol. 1365 © 2011 Materials Research Society
DOI: 10.1557/opl.2011.1384

Nonlinear Micro-Processing of Silicon by Ultrafast Fiber Laser at 1552 nm

Yoshiro Ito[1], Rie Tanabe[1] and Kozo Tada[2]
[1] Department of Mechanical Engineering, Nagaoka University of Technology, Kamitomioka, Nagaoka, Niigata 940-2188, Japan
[2] Citizen Finetech Miyota, 353, Yaehara, Tomi, Nagano 289-0406, Japan

ABSTRACT

Processing of transparent materials by non-linear absorption mechanisms induced by short pulse lasers has been applied in many fields. Silicon (Si) is widely used materials in microelectronics, MEMS and photonics. It is, however, not transparent for commonly used processing lasers in near infrared to ultraviolet spectral range and has not been a subject for the non-linear processing by lasers so far. In this paper, possibilities and capabilities of non-linear processing of Si by 900 fs, 1552nm laser radiation are described with special emphasis on application to frequency adjustment of a crystal oscillator in a package made from Si.

INTRODUCTION

Processing of transparent materials by non-linear absorption mechanisms induced by short pulse lasers has been applied in many fields such like photonics, inner marking and fiber optics. These applications are limited to transparent dielectrics and polymers. It is not applicable to metals and semiconductors because they are not transparent at commonly used laser wavelengths. Silicon (Si) is a widely used material in micro-electronics, MEMS and photonics. It is not transparent in near infrared to ultraviolet spectral range and has not been a target for the non-linear processing by lasers so far. However, semiconductors become transparent for photons energy of which are less than band gap energies of them.

In this paper, possibilities and capabilities of machining of Si by non-linear absorption mechanisms induced by an ultrafast, 1552 nm laser radiation are described with special emphasis on application to frequency adjustment of crystal oscillators in packages made from Si [1, 2, 3]. Band gap energy of Si corresponds to wavelength of 1128 nm and thus the laser light is expected to transmit the Si [4]. Machining at different positions, from the front surface of the first Si substrate to the back surface of the second substrate placed at the back of the first, is carried out in air. Focused outputs of the laser make trenches on the front surface of both Si substrates. Sharp and deep grooves are machined with some deposition on the edges on the front surface. When it is focused on the front surface of the second substrate placed at the back of the first Si substrate, trenches are made on the surface of the second with no detectable change on the first. When the laser is focused at the back surface, quite different machined features are observed on the back surface. A wide stripe of fine grained structure without any deep trench is formed on the back. Highly position selective machining of surfaces has been achieved by adjusting the focus to each surface with no detectable change on the other surfaces.

Gold (Au) film on the second substrate can be ablated without any change on front and back surfaces of the first substrate. Frequency adjustment of a crystal oscillator sealed in a Si package is tried and up-conversion of the frequency is achieved by removing small amount of thin gold film on the crystal with irradiation of the laser through the silicon package. Down-conversion of its frequency should be achieved by laser induced forward transfer (LIFT) of Au on the inner surfaces of Si package. Direct LIFT on Si has not been achieved yet but a thin SiO_2 layer among Si and Au makes it possible.

During the research, we have found that infrared laser light come through the silicon substrate is accompanied by green light.

EXPERIMENT

An ultra-fast fiber laser, which delivers 5 μJ, 900 fs pulses at 1552.6nm with repetition rate up to 500 kHz, was used as a processing laser. Output of the laser was focused by a 100X objective lens with focal length of 1.8 mm and NA of 0.85. Si samples were mounted on motor-driven stages and moved at 1 mm/s. Both p-type (20 Ω/cm) and high-impedance (50 kΩ/cm) silicon substrates of 300 μm thick were used. Au film of 120 nm thick with Cr buffer layer was deposited on surfaces of some substrates. Si substrates are transparent at the laser wavelength. Measured transmission spectrum of 300 μm thick p-type Si substrate is shown in figure 1. Transmittance for wavelength longer than 1200 nm is about 52 %, which is in good agreement with reflection losses at Si-air interfaces calculated from the refractive index of Si, 3.478 at 1550 nm. This large refractive index also caused large aberration when the light was focused from air into the Si substrate.

Two laser scanning methods employed in this study are shown in figure 2. Figure 2 (a) shows tilted scan path where two substrates are attached with spacers and set in tilted position against laser scan direction. Thus the focus position varies from the front surface of the substrate 1 to the back surface of substrate 2 along the scan path. Figure 2 (b) shows zigzag path where short line scans are repeated with changing the focus position stepwise.

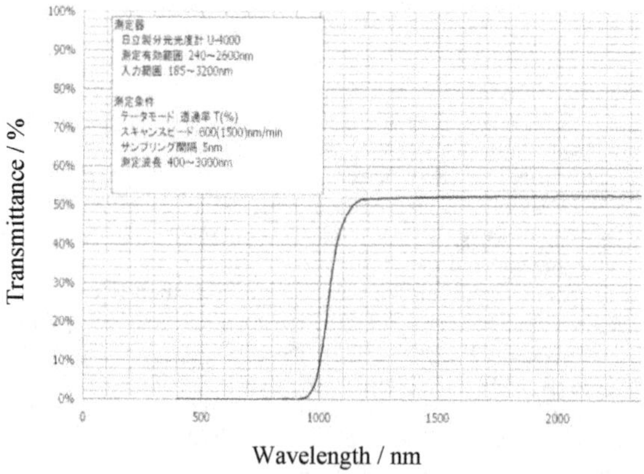

Figure 1. Transmittance spectrum of 300 μm thick p-type Si substrate measured by a spectro-photometer in air. Nearly constant transmittance for wavelength longer than 1200 nm indicates Si is a transparent material in this wavelength region. Measured spectrum of the high impedance Si does not show any difference.

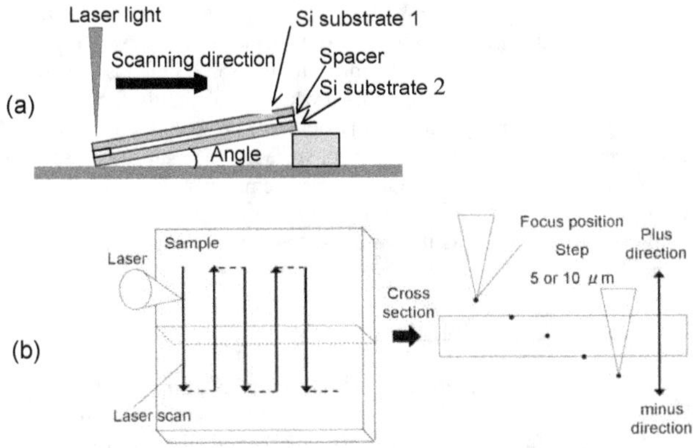

Figure 2. Schematic diagrams of laser irradiation path. (a) Tilted scan, (b) Zigzag scan.

RESULTS AND DISCUSSION

Micromachining of Si substrate

It has been demonstrated that infrared ultrafast laser can perform position-selective micromachining of Si substrate by non-linear absorption processes [2-3]. Tilted scan for two Si substrates, as indicated in figure 3 (a), showed that machining of narrow grooves on Si surfaces was achieved position selectively. When the focus fall on the front surfaces of two substrates grooves were formed on them. A groove was not formed when the focus position did not fall on the surface. Figure 3 (a) shows a scanning electron microscopy (SEM) image of the groove formed at the front surface of substrate 2. The laser was focused on the front surface of substrate 2 through the substrate 1, leaving no observable changes on both surfaces of substrate 1. If the

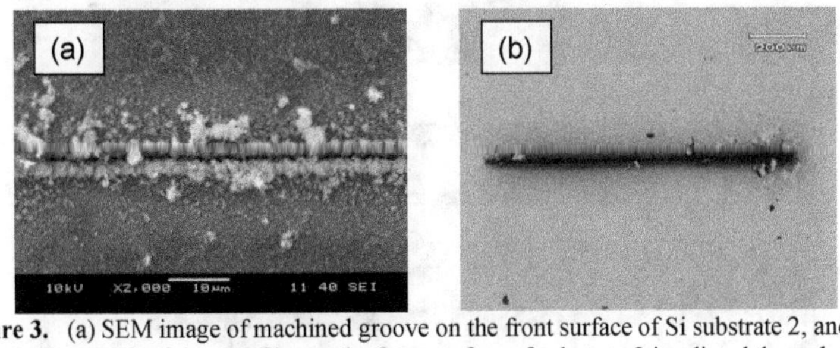

Figure 3. (a) SEM image of machined groove on the front surface of Si substrate 2, and (b) optical micrograph of thin Au film on the front surface of substrate 2 irradiated through substrate 1. Distance between two substrates was 64 μm. Laser pulse energy was 2.6 μJ, i.e., about 0.3 μJ on the sample, rep rate was 167 kHz and scan speed was 1 mm/s.

front surface of the substrate 2 was coated by Au, the Au film was ablated as shown in figure 3 (b). When the laser was focused on the substrate 1, only the front surface of the substrate 1 was machined with no change on the substrate 2. These results indicate that 1552.5 nm laser light normally transmits Si substrate while it can machine Si and Au when it was tightly focused.

We noted, however, that grooves were machined only on the front surfaces of Si and Au film on the front surfaces. Grooves were not machined on back surfaces of Si substrates and Au films on the back of them even if the focus position should fall on them. Only stripes of dark color were formed on the back surface of Si and no ablation occurred even for the Au films on it. Therefore, it is interesting and necessary to investigate nonlinear machining processes of Si in detail.

Machining of front and back surfaces of Si substrate

As shown in figure 3 (b), the laser was scanned by moving samples using a motor-driven x-y stage at 1mm/s. The focus position was moved stepwise for each line scan from above the top surface to below the back surface by moving an objective lens attached on a linear stage. Separation of each line scan was 100 μm. Figure 4 shows relative positions where laser induced changes appeared on the surfaces and optical micrographs of both front and back surfaces of the substrate irradiated by 1.8 μJ pulses at 500 kHz. Because of large refractive index of the Si, the focus positions inside the Si substrate shifted more than three times larger than the movement of the lens, 5 μm in this figure, with increased depth of focus. At focus positions abut + 90 μm to - 80 μm, laser induced surface changes were observed on the front surface. In addition to this area, a lesser change reappeared at -195 μm to -225 μm region. Considering the shift of focus position inside the Si substrate, the focus position should lie far below the back surface for this region. On the back surface, the laser induced changes were observed at -105 μm to -130 μm and at -310 μm to -320 μm. Si has very large refractive index, 3.478 at around 1550 nm, and thus it is estimated that the focus position should lie close to the back surface when the lens is moved at

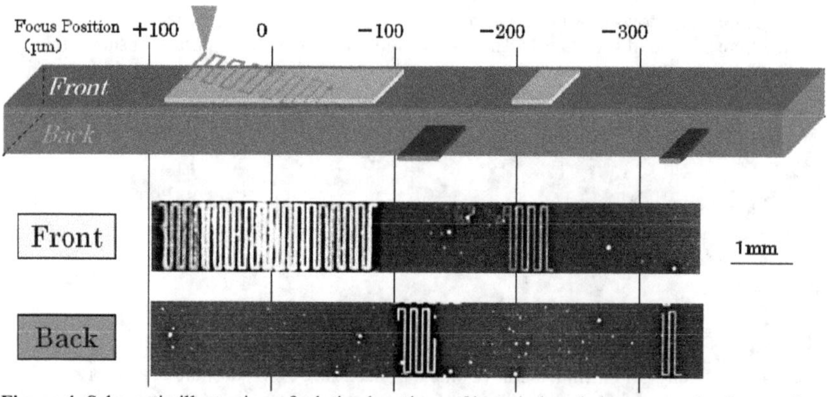

Figure 4. Schematic illustration of relative locations of laser induced changes on the front and the back surfaces of Si substrate (top) and optical micrographs of the front surface (middle) and the back surface (bottom). Laser pulse energy was 1.8 μJ and repetition rate was 500 kHz.

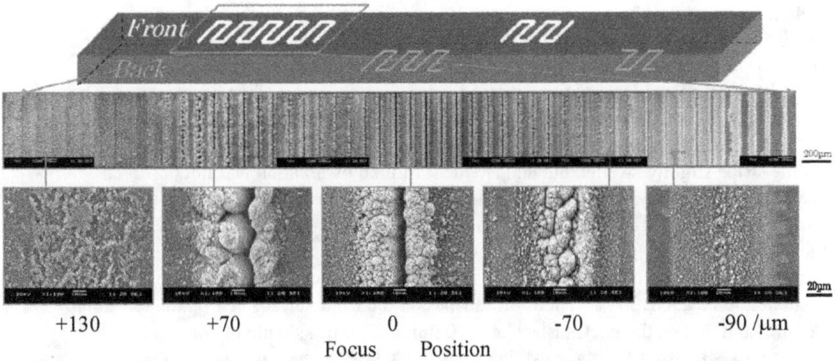

Figure 5. SEM images of machined structures on the front surface. Laser pulse energy was 18.5 μJ with 100 kHz repetition rate and thus larger area was machined compared to figure 4.

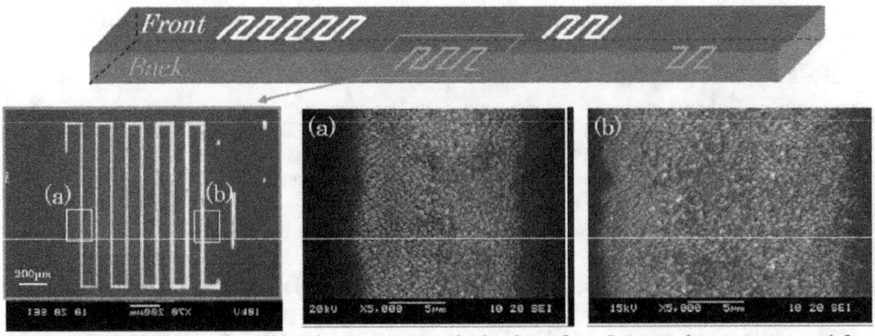

Figure 6. SEM images of machined structures on the back surface. Laser pulse energy was 1.8 μJ with 500 kHz repetition rate. (a) and (b) indicate magnified images of corresponding positions.

-100 μm position. Groove structures were formed only for focus positions between +50 to -50 μm, where laser fluence should be high enough to induce ablation by non-linear mechanisms. Stripes of roughened surface structure were formed on the back surface for focus positions at -105 μm to -130 μm and -310 μm to -320 μm. The same stripes were found on the front surface at -195 μm to -225 μm region.

Figure 5 shows SEM images of stripes and grooves on the front surface of Si formed by zigzag scan of the laser. The distance between the lens and the sample surface decreased 5μm for each scan. Higher pulse energy of 18.5 μJ was used. Therefore, laser induced changes were left in wider area, and thus wider focus position region, compared with figure 4. When the laser was defocused, irradiated area looked as darkened compared with non-irradiated part and the SEM images in the figure suggested that the change was due to roughening of the surface. Formation of grooves was observed at the centers of irradiated paths for +80 to -70 μm and some thermal effects such as deposition of debris were observed around the grooves. Therefore, it is evident that laser ablation is induced by non-linear absorption processes on Si surface. For large defocusing distances, width of the stripe decreased and finally disappeared.

Figure 6 shows SEM images of fabricated structures on the back surface. Laser pulse energy was 1.8 μJ with 500 kHz repetition rate. Stripes of roughened surface were formed. The stripe consists of nanometer size dots. Width of the stripe is about 15 μm at -105 μm, increased to about 22 μm at -145 μm and disappeared. Measurements by a laser microscope indicated that the area of the stripe slightly swelled out and any dent formed by ablation was not observed. Even if the laser pulse energy was increased, it was not possible to ablate the back surface of Si substrate. Additional stripe structures formed on the front surfaces for -195 μm to -225 μm and the back for -310 μm to -320 μm positions revealed the same fine structure to this.

Amount of reflection at Si-air interface is rather large due to the large refractive index of Si and calculated to be about 30 %. Therefore, stripes on the front surface at around -195 μm to -225 μm region and that on the back surface at -310 μm to -320 μm should be formed by reflected laser pulses at Si-air interface. At these positions, laser came out from Si to air and formed similar roughened surface structure. If this is true, the last stripes on the back surface were formed by very small energy pulses, less than 3% of initial incident energy. It is interesting to note that when the laser hit the surface of Si from air, the groove structure is formed, and thus the laser ablation occurs, even by the smaller laser energy as indicated in figure 3.

Changes observed in Si substrate

In the experiment shown in figure 4, there were regions where no surface change was occurred. Laser pulse should be focused inside the substrate in these region and laser fluences on the front or back surfaces were lower than the damage threshold values. Does the absorption of the laser pulse occur inside Si and cause any structural change? It is suggested that when the visible or near ir nanosecond laser pulses are focused at inside of Si substrate, the laser could induce small cracks in Si, which make possible to dice the substrate [5, 6]. The proposed mechanism for this is that initial temperature rise around the focus point due to small linear absorption causes the increase of absorption, thus a positive feedback process starts and enough energy is deposited to form micro-cracks [7, 8]. In our case, however, laser photon energy is less than the band gap energy of Si and the linear absorption does not occur. In addition, pulse width is very short and we may neglect any feedback mechanism due to local temperature rise. Top picture in figure 7 shows optical micrograph of the cross section of the Si substrate. The sample was cut in perpendicular direction to the zigzag path and polished. Grooves on the front surface were clearly observed. Their depth deepened as the focus position approach to the surface and became deepest of 60 μm when it was on the surface. No other change was observed inside the substrate or in the vicinity of the back surface. The sample was etched by KOH solution and the result was shown in the middle of the figure 7. The grooves on the front surface become clearer. In addition, black lines appear in the vicinity of the back surface. Bottom picture in the figure shows dark stripes observed on the back surface at the same position. The positions of black lines exactly coincide with those of the stripes on the back surface. No other change is observed inside the substrate even in the etched sample. Detailed SEM observations of dark lines appeared in the vicinity and on the back surface suggest that original single crystal structure changed into multi-grain structures smaller than one micrometer.

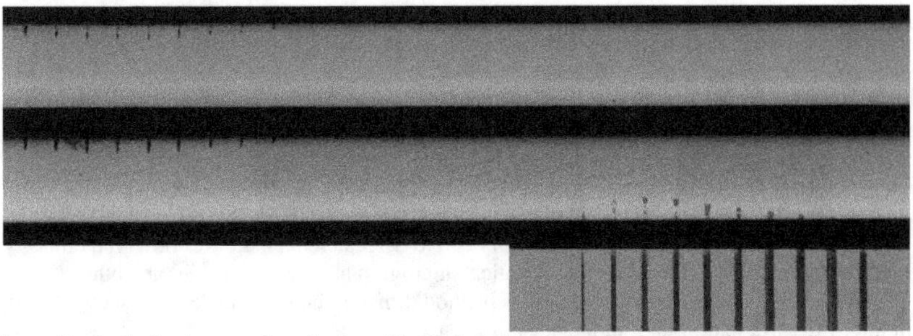

Figure 7. Optical micrographs of p-type Si substrate irradiated in zigzag path. Top: Cut and polished cross section. Middle: Etched by KOH solution after polishing. Bottom: Stripes formed on the back surface. Thickness of the substrate is 300 μm. Similar results are obtained for high impedance Si substrate.

These results imply that threshold energy fluences for ablation or structure change on the surface of Si are much smaller than that for modification of bulk Si. The high refractive index of Si should cause strong aberration on focusing light in Si substrate, resulted in defocusing effect and thus energy fluence inside Si might be decreased compared to that at the front surface. At the back surface, however, focusing situation should be the same in the bulk. Once the surface change occur, structure changes seem to propagate toward inside, as shown in the figure but no other change independent of the surface change is observed. Thus it is clear that the threshold fluence of bulk Si is much larger than those for Si surfaces. These results are not affected by conductivity of the Si substrate.

Frequency adjustment of crystal oscillator in Si package

A versatile technique to adjust the frequency of the crystal oscillator should be capable to accomplish both up- and down-conversion of the frequency. The up-conversion is achieved by decreasing the mass of the fork-part of the crystal while down-conversion by increasing the mass by some laser induced processes such as LIFT. Amount of frequency change is represented by ppm from its designed value of 32.4700 kHz.

Up-conversion of frequency: ablation of Au film through Si substrate

Up-conversion of the frequency should be achieved by removal of Au film deposited on the crystal surface. Laser ablation of a thin Au film on the second Si substrate by laser irradiation through the Si substrate has already been demonstrated in previous section. Accuracy and robustness of the process, however, have to be tested. Custom-made sample holder was used to test commercial Si-packaged crystal oscillator. This holder had a metal mask which allowed us to limit the laser irradiation area to the part of crystal fork where the Au thin film was deposited. In addition, it was able to measure the frequency of the crystal during and immediately after the laser irradiation. During the laser irradiation, however, measured frequency was not stable due to the laser induced shock and thermal effects. The frequency became measurable immediately after the laser irradiation and remained stable after the irradiation. Figure 8 shows infrared optical micrographs of the sample after laser irradiation. Three parallel scans were carried out with 2.6 μJ pulses and the traces of scans are seen as three black lines in dotted square on the head. Focus position was estimated to be placed on the crystal surface, but was not adjusted precisely, because the laser irradiation system and the holder were not integrated with an infrared microscope and thus we could not 'see' inside the package during the experiment. Measured shift of the frequency was about 25 ppm per scan in average and satisfied the required accuracy in frequency adjustment process. In the figure 8 (b), dark, cloud-like shade is observed around the three lines. Detailed observation revealed that this dark cloud was Au ablated from the crystal surface and deposited on the back surface of the Si lid. No detectable damage was observed on the top surface of the lid by both visible and ir optical micrographs. The frequencies and other electric parameters of the oscillators tuned by this method turned to be very stable: they showed little drift after laser irradiation for 8 months.

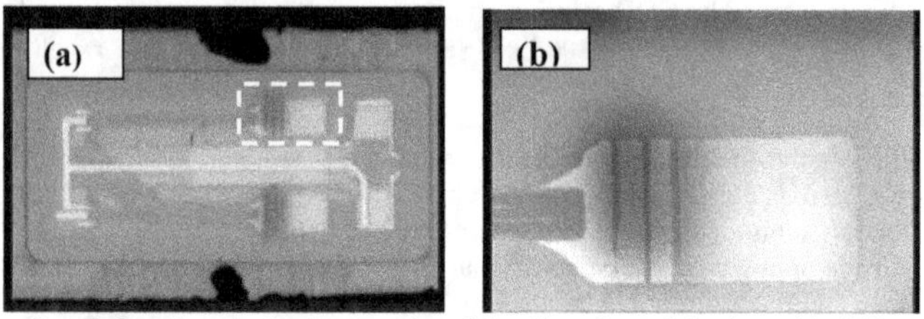

Figure 8. Infrared optical micrographs of the Si packaged device with different magnifications. (a) whole package, (b) magnified picture of laser irradiated part. The size of the device (a) is 2 mm x 1.6 mm. Gold thin film of 120 nm was spattered on the part of tuning fork indicated by dotted line. The vertical lines inside the dotted square are machined groove by the laser irradiation. (b) shows the magnified image of the irradiated part.

<u>Down-conversion of frequency: LIFT through Si substrate</u>

To decrease the crystal frequency, small mass should be added to the crystal. One of the possible processes laser induced forward transfer (LIFT) of Au film on the back surface of the Si lid onto the crystal fork below. In this way, both up- and down-conversion should be achieved by changing the focus position of the infrared laser pulse and irradiated positions. The laser should be capable to achieve LIFT of Au film because the ablation of Au on the surface of the second Si substrate located at the back of Si substrate successfully, as shown in figure 3. To make this certain, LIFT of Au on Si, SiO_2 and sintered Al_2O_3 substrates was carried out. LIFT of Au film was successfully achieved for SiO_2 and Al_2O_3 substrates. Au film on the back of Si substrate was, however, not ablated, as shown in figure 9. In the figure, LIFT results by 1 μJ laser pulses for Si and SiO_2 substrates are shown. Laser should be focused at the back of substrates. Au film on SiO_2 substrate was ablated without any change on the front surface and deposition of Au on acceptor surface was evident. In case of Si substrate, however, we could not find any detectable change on the front surface of Si or Au film on the back. The back surface of Si substrate suffers the morphology change by laser irradiation as shown in the previous section under the same irradiation conditions. This suggests that the threshold fluence of Si surface machining is less than that for Au ablation, even though Si machining would be non-linear processes. To clarify this situation, Au film was sputtered in small rectangular patterns on the back surface of Si

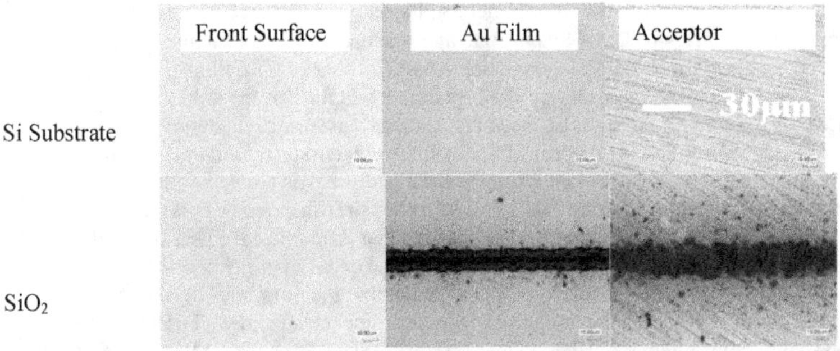

Figure 9. Optical micrographs showing the LIFT of Au films on Si (top row) and SiO_2 (bottom row) substrates to Si substrates. Laser pulse energy is 1 μJ. LIFT is achieved only for SiO_2 substrate and any change is observed for Si substrate.

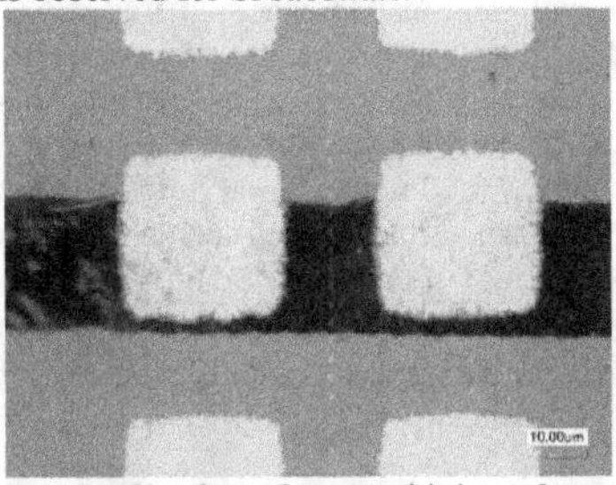

Figure 10. Optical micrograph of back surface machining of patterned Au film and Si. Laser was focused at the back of the substrate and scanned. Surface structure change of Si is evident as black stripe at the middle of the picture but square shaped gold films of 120 nm thick do not show any change even on the stripe.

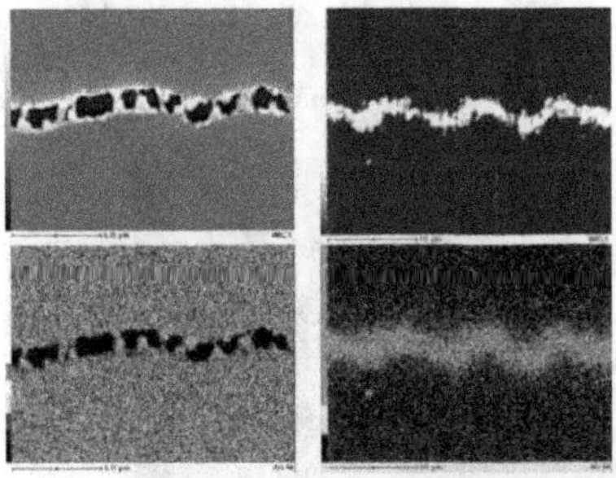

Figure 11. SEM images of Au film on the back of Si/SiO_2 substrate (top left) and deposition of Au on the front surface of acceptor substrate (top right). EDS mapping results of Au are shown below the corresponding SEM images. Thickness of SiO_2 layer on the back of Si is 10 μm and laser pulse energy is 1 μJ.

substrate and laser was irradiated from the front. The result is shown in figure 10. Laser pulses of 2 μJ were focused at the back surface of Si substrate and scanned. Surface structure change of Si is evident as black stripe at the middle of the picture but square shaped gold films of 120 nm thick do not show any change even on the stripe. To investigate this further, the ablation threshold fluencies for Au film on the front surface of Si and Si front surface itself were compared under direct laser irradiation. It is confirmed that the ablation threshold of Au film on Si is much greater than that of Si surface even when the laser is directly irradiated on them. Our results indicate that the damage threshold of Si surface is much lower than Au or bulk Si. There are wide ranges of studies on Si surfaces and it is well known that properties of Si surfaces differ from those of bulk Si. Chemical reactions with species in air also occur in our experimental condition. Thus we would expect that changes in optical and electronic properties of Si surface might reduce the threshold. Most of these studies are, however, for 'clean and well defined' Si surfaces in high vacuum environment. Further studies on 'dirty' Si surfaces would be required to understand the reasons for the low damage threshold of Si surfaces.

How can we achieve LIFT of Au on the back of Si lid? We tried to modify the surface of Si. Thin layer of SiO_2 was sputtered on Si and Au film was sputtered on SiO_2 layer. Thus Si/SiO_2/Au structure was made and the laser was irradiated to Au through Si substrate. Results are shown in figure 11. In the figure, the SEM images of Au film on the back of Si/SiO_2 substrate and deposition of Au on the acceptor substrate are shown with corresponding EDS mapping of Au. Thickness of SiO_2 layer was 10 μm and laser pulse energy was1 μJ. It is clear that LIFT of Au film on Si/SiO_2 surface is possible. There is no damage on the front surface of the donor and the acceptor Si substrates. SiO_2 layers of 10, 5 and 1 μm thick were made on Si and Au film on them was ablated through Si. It is found that 1 μm thick SiO_2 is enough to achieve the LIFT of Au film. The results provide us a possible technique to achieve the down-conversion of the oscillator frequency through Si lid.

Generation of green light through Si substrate

During these experiments, we found that green light was visible at the back of the Si substrate, as shown in figure 12. It was rather dim and only seen in the dark. Figure 11 shows the green light on a screen placed at the back of the Si substrate. The 1552 nm laser pulses with 100 kHz repetition were focused from front surface to inside Si. The spectrum of this green light was measured by a multichannel spectrometer. The obtained spectrum is a line spectrum at 516

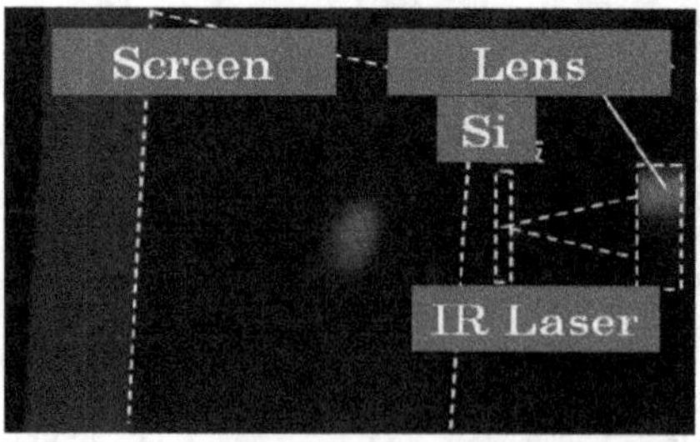

Figure 12. Emission of green light through Si substrate.

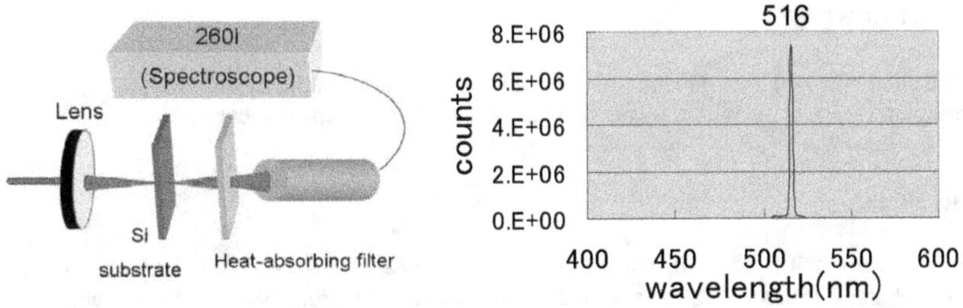

Figure 13. Spectral measurement arrangement and obtained spectrum of green emission through Si. 1552 nm laser was illuminated from the left of Si and the spectrum was measured at the back of the Si substrate.

nm and shown in figure 13. The power of green light was very low and was not measurable so far. Most of laser power was transmitted as original infrared wavelength light. Part of the infrared radiation seems to be converted to the green. Origin of this green light is, however, not clear yet. It is interesting that its wavelength is very close to one third of original ir wavelength.

CONCLUSIONS

Capabilities of machining of Si by non-linear absorption mechanisms induced by the ultrafast, 1552 nm laser pulses has been demonstrated. Highly position selective machining of a silicon substrate and machining of another substance placed at the back of it are successfully achieved. Narrow grooves have been machined on the front surfaces of both the first and the second Si substrates placed with small gap by focusing the laser radiation at each position without any damage on other parts. On the other hand, the back surfaces of Si substrates are not ablated. The gold film deposited on the second substrate placed at the back of the first Si substrate is precisely removed by the laser irradiation thorough the first Si substrate. The threshold fluence for surface ablation or modification is far smaller than that for modification of bulk Si.

The method is applied to the frequency adjustment of the crystal oscillator sealed in the Si-made package. Using the specially designed sample holder, the increase of the frequency of the crystal oscillator has been achieved by removing the gold film deposited on the crystal surface by laser irradiation through Si lid. This technique is proved to have enough accuracy applicable to adjustment of the oscillator sealed in the Si package. To decrease its frequency, however, LIFT of Au on the inner surface of the package must be carried out. So far, this is achieved by making a thin SiO_2 layer between Si and Au. This would be applicable to other tuning processes of many MEMS devices made from silicon or other semiconductors.

To utilize the high position selectivity of non-linear processes for Si, it is important to compensate the aberration of laser light due to its high refractive index. It is also important to integrate both a laser machining system and an ir microscope.

ACKNOWLEDGMENTS

The authors would like to thank Messrs. Sakurai, S. Nemoto and K. Hoshi for their support in experiment. Research supported by the A-STEP program of the Japan Science and Technology Agency.

REFERENCES

1. Japan Patent 2009-124310, 2009-246188
2. Y. Ito, F. Sato, Y. Shinohe, R. Tanabe, and K. Tada, Proc. SPIE Vol. **7584**, 75840M (Feb. 17, 2010)
3. R. Tanabe, F. Sato, Y. Shinohe, K. Tada and Y. Ito, Proc of LPM2010, #10-25 (2010) http://www.jlps.gr.jp/en/proc/lpm/10/
4. See for example, M. Shur, 'Physics of Semiconductor Devices', Prentice-Hall Inc., New Jersy, (1990) p478
5. K. Atsumi: The Japan Society of Mechanical Engineers, No. 1068, Vol. **110**, 14-16 (2007) (in Japanese)
6. F. Fukuyo, K. Fukumitsu, and N. Uchiyama, "The stealth dicing technologies and their application", Proceedings of LPM 2005 (2005)
7. K. Fukumitsu, M. Kumagai, E. Ohmura, H. Morita, K. Atsumi and N. Uchiyama, "The mechanism of semiconductor wafer dicing by stealth dicing technology", Proceedings of LAMP2006 (2006)
8. E. Ohmura, M. Kumagai, K. Fukumitsu, M. Nakano, N. Uchiyama and H. Morita, "Internal modification of ultra thin silicon wafer by permiable pulse laser", Journal of the Japan Society for Precision Engineering, **74** (3), 275-281, (2008), and references therein. (in Japanese)

Mater. Res. Soc. Symp. Proc. Vol. 1365 © 2011 Materials Research Society
DOI: 10.1557/opl.2011.887

Ultrafast Laser Processing of Hybrid Micro- and Nano-structures in Silicate Glasses

Pavel Mardilovich[1], Luke Fletcher[2], Neil Troy[2], Lihmei Yang[3], Huan Huang[3], Subhash Risbud[1] and Denise M. Krol[2]
[1]Department of Chemical Engineering and Materials Science, University of California, Davis
One Shields Ave, Davis, CA 95616, U.S.A.
[2]Department of Applied Science, University of California, Davis
One Shields Ave, Davis, CA 95616, U.S.A.
[3]PolarOnyx Inc., 2526 Qume Drive, Suites 17 & 18, San Jose, CA 95131

ABSTRACT

This study describes the fabrication of hybrid micro- and nanostructures of semiconductor nanocrystals arranged in microscopic lines inside of a borosilicate glass doped with CdS_xSe_{1-x}. This was performed using a two step process of (1) ultrafast laser modification and (2) heat treatment. The glass was photomodified using focused sub-picosecond infra-red pulses with 1 MHz repetition rate to create linear domains with local compositional variations. Heat treating the sample at temperatures near glass transition preferentially precipitated semiconductor in the modified regions, as evidenced by confocal fluorescence microscopy. The optical properties of the precipitated nanocrystals varied with heat treatment duration.

INTRODUCTION

Ultrafast laser processing of glasses has potential for applications in a host of photonic devices [1]. Short pulses, in the femtosecond to picosecond range, when focused inside a normally transparent glass produce extremely high electromagnetic fields at the focal point, which leads to non-linear effects and absorption of laser energy. The rapid absorption and dissipation of energy leads to structural changes [2], which can be induced with a great degree of control in three dimensions. This micron scale modification lies at the heart of a lot of currently investigated photonic devices, such as basic waveguides [3,4] interferometers [5], couples and splitters [6], etc..

Recently there has been interest and notable advances in using active glasses, i.e. glasses that can exhibit signal gain [7], and that can be used either for signal amplification along the waveguide, or, with proper cavity construction, compact waveguide lasers [8]. The focus of active glass waveguides so far has been on glasses doped with ions, such as Er and Yr. One shortcoming of such systems, is that in rare earth ions optical transitions, dictating gain bandwidth, are narrow and do not easily yield themselves to tuning.

Previously, quantum dots (QDs) have been considered as good candidates for gain media due to being easily invertible, thus having low lasing threshold, and, perhaps more importantly, having optical properties tunable with particle size [9]. In this light, quantum dot doped glasses make intriguing candidates for the aforementioned photonic applications, as the particle size in these glasses, and therefore their optical properties, can be easily controlled by adjusting simple processing conditions that govern their nucleation and growth, such as heat treatment temperature and duration [10, 11].

To explore this possibility we aim to create linear domains of quantum dots, defined by ultrafast laser exposure. We use a two step process: an ultrafast laser modification followed by heat treatment.

EXPERIMENT

A commercially available colloidal filter glass was used as a starting material. In this case, OG570 filter glass from Schott Inc was used. This glass is doped with CdS_xSe_{1-x}, which is already precipitated, and is responsible for the filter's color. To dissolve the semiconductor crystals, the OG570 glass was melted at 1450 °C for ½ hour, then cast in to a graphite mold, and annealed at 510 °C, its glass transition temperature, to remove internal stresses. This approach is similar to what was used in a number of earlier studies on precipitation dynamics of semiconductor in glass [12] The result was a clear glass, hereafter referred to as OG570-Q, which was then cut and polished into workable samples.

The ultrafast laser modification of the glass was performed at using a Uranus series fiber laser from PolarOnyx Inc.. The laser wavelength was 1030 nm, with pulse width <1 ps, and repetition rate of 1 MHz. After brief preliminary experimentation it was found that pulse energies of 300 nJ and above produced the desired scale of modification, and 300 nJ was the pulse energy used in this experiment. The laser beam was focused with a 50X objective, with N.A. = 0.55. The sample was translated transversely with respect to the beam focus at a rate of 100 μm/s to produce a line consisting of a series of overlapping modified regions.

To expose the linear modification for analysis, the irradiated sample was cut into cross sections and polished. The cross sections were analyzed using scanning electron microscopy (SEM) and wavelength dispersive spectroscopy (WDS). Then the samples were subjected to heat treatment at 500 °C to precipitate the quantum dots. The progress was monitored using confocal fluorescence microscopy in a setup detailed in ref. 4, using a diode 473 nm excitation laser and a 20X focusing objective with N.A. = 0.4. The fluorescence spectra were acquired from a raster of points in a 50 micron by 50 micron area at 1 micron steps. The analysis of the fluorescence spectra consisted of the mapping of integrated spectral intensities for all the points to produce an image of the analyzed area, as well as a closer analysis of the individual spectra.

DISCUSSION

Figure 1 shows a series of six lines, spaced 10 microns apart, "written" in the OG570-Q sample. Figure 1a shows the view of modifications from the top. It is apparent that each line consists of overlapping spherical modifications, with the overlap direction corresponding to the back-and-forth translation direction of the sample. It's important to note that at 1 MHz pulse repetition rate and 100 μm/s translation speed the laser pulse would hit the sample 10^4 times for each micron traveled, so the spheres that we see in figure 1a are certainly not the result of individual pulses.

The sample cross section that is shown in figure 1b shows very apparent modifications. The slightly zigzagging pattern of the locations of the lines is due to a small hysteresis of the air-bearing translation stage as the sample was moved back and forth to write the lines.

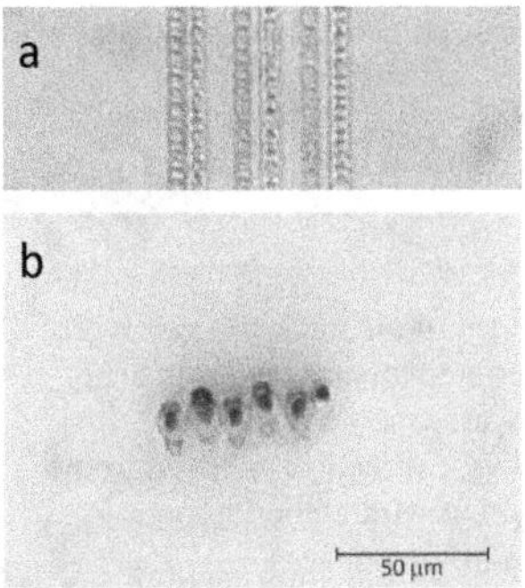

Figure 1. Ultrafast laser-induced linear modifications in OG570-Q glass; view of lines from the top (a) and of their cross section (b). The direction of the modifying laser is into the page in image (a), and from the top in plane with the page in (b).

Elemental redistribution

SEM backscattered electron (BSE) image is shown in figure 2. The contrast in BSE imaging is due to different elastic scattering cross sections of different elements, with heavier elements scattering more and showing a brighter signal. From this it is apparent that photomodification of OG570-Q has resulted in desegregation of the elements, a result observed previously [13].

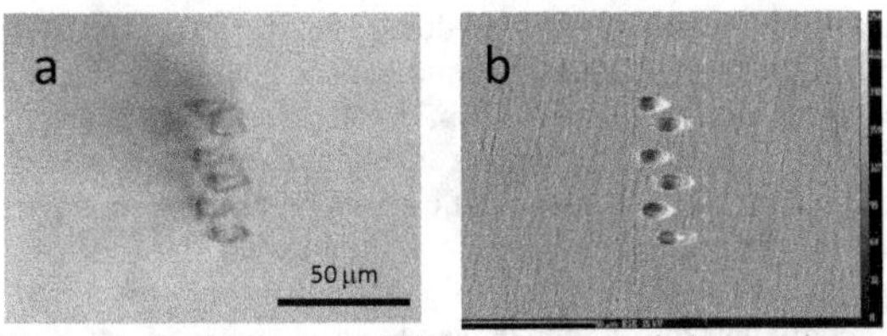

Figure 2. Optical (a) and SEM backscattered electron image (b) of the modified lines' cross-sections.

Wavelength dispersive spectroscopy gives a closer look at the distribution of major cations around the modified region (figure 3). From literature [ref. thermophoresis] it is expected for network modifiers to migrate to the outside of the modified region, while the more strongly bonded network formers generally stay towards the center in what is known as the Ludwig-Soret effect. This is indeed what we see here, with silicon, part of the silica network, showing a higher concentration in the center of modification (fig. 3c), and sodium, potassium and zinc, cations of network modifying oxides, migrating to the periphery of the modified regions (fig 3d-f).

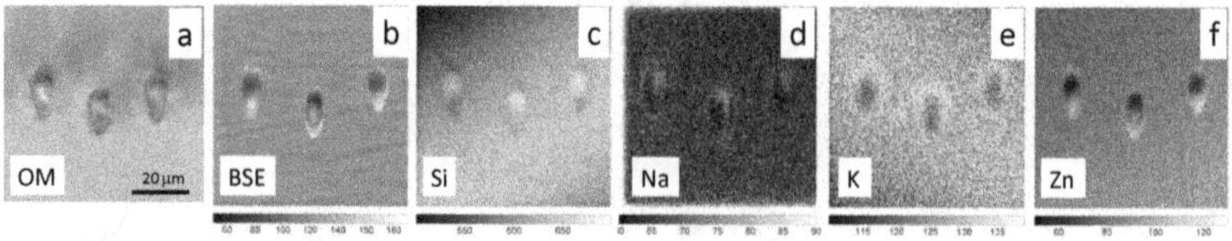

Figure 3. Optical micrograph (a) and backscattered electron image (b) of an area analyzed with wavelength dispersive spectroscopy. Silicon is more highly concentrated towards the center of the modification (c), sodium and potassium (d,e) are distributed towards the periphery with slightly higher concentrations towards the top of the modification in the direction opposite to the modifying beam (from the top); zinc, likewise, moved to periphery, but with higher amount moving in the direction of the modifying beam, towards the bottom of the image.

However, it is interesting to note that zinc and potassium display very different distributions. The signal of sodium in WDS is very low due to its tendency to diffuse under the probe's electron beam [14], but it appears to follow generally the same pattern as potassium. This indicates that there may be fundamental differences in diffusion mechanics between the cations, with cation valence possibly playing a role.

From figure 3 it can be seen that as a result of photomodification the cross section of the line forms roughly three chemically distinct regions: K-rich, Si-rich, and Zn-rich in the direction of the modifying beam.

Fluorescence

Figure 4 shows evolution of fluorescence (fig 4b-g) of the modified area (fig 4a) as the sample is treated at 500 °C. Initially there is little fluorescence of the modifications themselves, and a degree of fluorescence in the area immediately surrounding them (fig 4b). However, as the sample undergoes heat treatment, the fluorescence of the surrounding area disappears, while the modifications develop brighter and brighter fluorescence in areas that correspond to increased potassium and zinc concentrations. The rapidly disappearing fluorescence in the area surrounding the modifications is likely due to defects in the glass, which are commonly produced in ultrafast laser modification, and are generally unstable at high temperatures [15].

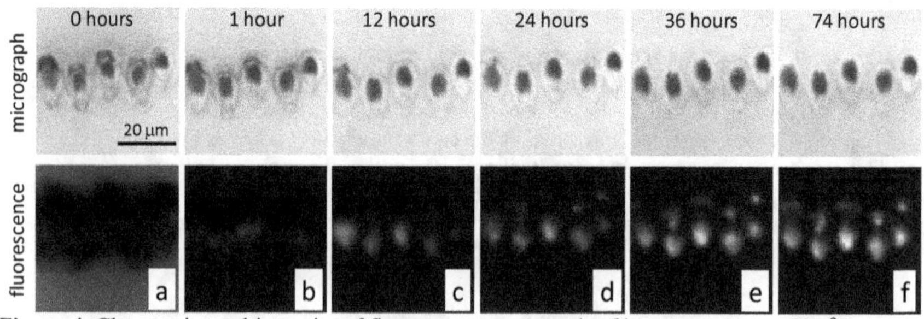

Figure 4. Changes in total intensity of fluorescence as a result of heat treatment at 500 °C; untreated (a), 1 hour (b), 12 hours (c), 24 hours (d), 36 hours (e) and 74 hours (f).

Fluorescence spectra from a sample point in the K-rich zone are shown in Figure 5. Here we can see that after 12 hours of heat treatment a distinct peak emerges at 590 nm (fig. 5c). With subsequent exposure to the elevated temperature, the peak grows in intensity and shifts right toward the redder wavelengths (fig 5d-f). This is typical of quantum dot growth, as characteristic optical transitions shift from blue to red with increasing particle size. For comparison, the fluorescence spectra of the area immediately surrounding the modification is shown in Figure 6, where we see little change after the first hour (fig 6b-f), indicating that either nothing happens, of that the precipitating semiconductor is too small to be resonant with the 473 nm excitation beam.

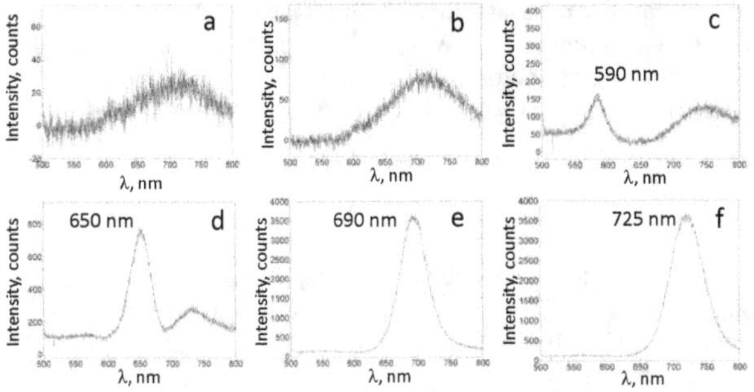

Figure 5. Evolution of fluorescence at a sample point in a potassium rich region of the modifications at zero (a), 1 hour (b), 12 hours (c), 24 hours (d), 36 hours (e) and 74 hours (f).

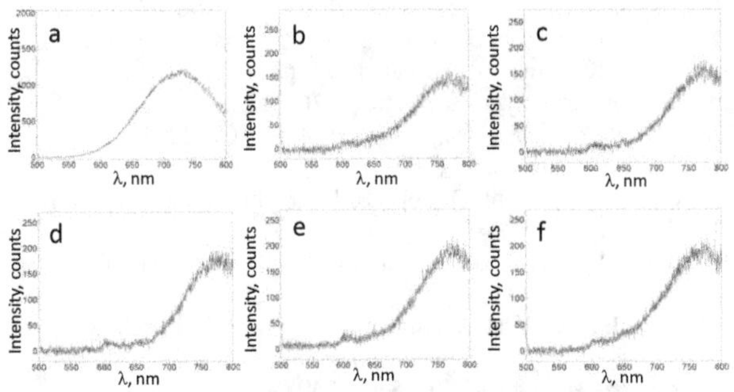

Figure 6. Evolution of fluorescence at a sample point in a region immediately surrounding the modifications at zero (a), 1 hour (b), 12 hours (c), 24 hours (d), 36 hours (e) and 74 hours (f).

Mechanism

The presence of network modifiers, such as potassium or zinc, serves to break up the borosilicate glass network, generally depressing the melting and glass transition temperatures. From this we can surmise that the glass transition in K-rich and Zn-rich regions occurs at a lower temperature than in the bulk, and for any given temperature, the network mobility is considerably higher in these regions than in the rest of the glass. As a consequence, at a particular range of temperatures, it should be possible to obtain a reasonably fast semiconductor nucleation and growth in the irradiated areas, while keeping precipitation extremely slow in the surrounding bulk. Of course, this ignores the changes to the solubility limit of the semiconductor components due to local compositional changes as well as possible phase separation of new compositions. These factors need to be taken into account if developing a more comprehensive model.

CONCLUSIONS

Using a high repetition rate ultrafast laser, a glass doped with dissolved CdS_xSe_{1-x} was modified to produce linear domains a few microns across. WDS study of cross sections indicated elemental desegregation with three distinct regions across those lines: K-rich, Si-rich and Zn-rich. Of these, the potassium and zinc rich areas displayed changes in fluorescence with heat treatment characteristic of growing quantum dots.

ACKNOWLEDGMENTS

The authors of this paper would like to acknowledge the financial support of this research from National Science Foundation (NSF) grant no. CMMI-0825572 and Air Force Office of Scientific Research (AFOSR) grant no. FA9550-10-C-0001.

REFERENCES

1. G. Della Valle, R. Osellame and P. Laporta, *Journal of Optics A: Pure and Applied Optics* **11**, 1 (2009).
2. C.B. Schaffer, A. Brodeur, E. Mazur, *Measurement Science and Technology* **12**, 1784 (2001).
3. K.M. Davis, K. Miura, N. Sugimoto and K. Hirao, *Optics Letters* **21**, pp. 1729 (1996).
4. J.W. Chan, T. Huser, S. Risbud and D.M. Krol, *Applied Physics A* **76**, 367 (2003).
5. S. Nolte, M. Will, J Burghoff and A. Tennermann, *Applied Physics A* **77**, 109 (2003).
6. A. M. Streltsov and N. F. Borelli, *Optics Letters* **26**, 42 (2001).
7. R. Osellame, S. Taccheo, G. Cerullo, M. Marangoni, D. Polli, R. Ramponi, P. Laporta and S. De Silvestri, *Electronic Letters* **38** 964 (2002).
8. R. Osellame, G. Della Valle, N. Chiodo, S. Taccheo, P. Laporta, O. Svelto and G. Gerullo, *Applied Physics A* **93,** 17 (2008).
9. J.M. Auxier, A Schulzgen, M.M. Morell, B.R. West, S. Honkanen, S. Sen, N.F. Borelli and N. Peyghambarian, *Fiber Lasers II: Technology, Systems and Applications, Edited by L.N. Durvasula, A.J.W. Brown and J Nilsson, Proceedings of SPIE* **5709**, 249 (2005).
10. L.-C. Liu and S. Risbud, *Journal of Applied Physics* **68**, 28 (1990).
11. A. Ekimov, *Journal of Luminescence* **70**, 1 (1996).
12. N.F. Borelli, D.W. Hall, H.J. Holland and D.W. Smith, *Journal of Applied Physics* **61**, 5398 (1987).
13. Y Liu, M. Shimizu, B. Zhu, Y. Dai, B. Quian, J. Qui, Y. Shimotsuma, K. Miura and K. Hirao, *Optics Letters* **34**, 136 (2009).
14. M.P. Borom and R.E. Hanneman, *Journa of Applied Physics* **38**, 2406 (1967).
15. A. Salimina, R Vallee, S.L. Chin, *Optics Communications* **256**, 422 (2005).

Mater. Res. Soc. Symp. Proc. Vol. 1365 © 2011 Materials Research Society
DOI: 10.1557/opl.2011.1318

Ultra-fast photoluminescence in fused silica surface flaws susceptible to laser damage

Ted A. Laurence[1], Jeff D. Bude[1], and Nan Shen[1]
[1]Condensed Matter and Materials Division, Lawrence Livermore National Laboratory
Livermore, CA 94550, U.S.A.

ABSTRACT

Using high-sensitivity confocal time-resolved photoluminescence (PL) techniques, we found an ultrafast PL (40 ps-5 ns) from impurity-free surface flaws on fused silica. This PL is excited by the single-photon absorption of sub-band gap light. Regions which exhibit this PL are strongly absorptive well below the band gap, as evidenced by a propensity to damage with 3.5 eV nanosecond-scale laser pulses. Very high defect densities are needed to explain the damage thresholds observed. For such high defect densities, significant interactions between defects may strongly affect the temporal characteristics of the emission of electronic excitations. We propose that the distribution in lifetimes observed is not simply due to a large variety of defect states, but due to a variety of energy transfer interactions between defect states.

INTRODUCTION

The goal of the National Ignition Facility at Lawrence Livermore National Laboratory is to obtain nuclear fusion in a laboratory setting using laser-based inertial confinement fusion. Laser pulses from 192 beams implode a fuel capsule, increasing the temperature and pressure to the regime where nuclear fusion can occur. Such high energy laser pulses can cause damage in the final focusing optics. This laser damage from nanosecond-scale pulses occurs in a regime with important differences from femtosecond laser damage. For nanosecond scale pulses, the electronic excitations have sufficient time to transfer their energy to the lattice [1]. This fact plays a key role in the development of laser-assisted absorption fronts [2], which lead to the large damage sites caused by temperature-activated absorption. Another important point is that, for the ns-scale pulses at the focusing optics, the intensities in pulses for which damage occurs is only near 2 GW/cm^2, which is much lower than intensities at which multiphoton absorption $\chi^{(3)}$ processes are observed. In order to explain this laser damage, high defect densities are required. We previously estimated that absorptivities of at least α=1000 cm^{-1} are required to explain laser damage of silica surface flaws [3]. For example, if the absorption cross-section σ of a defect is very high at 10^{-16} cm^{-2}, the defect density required is $n = \alpha / \sigma = 10^{19}$ cm^{-3}. For smaller cross-sections, n would have to be greater.

We recently discovered a novel photoluminescence with distributed fluorescence lifetimes in surface flaws in fused silica that are associated with laser damage by high fluence, ns-scale lasers [3]. This PL exhibits a broad distribution in lifetimes and emission and excitation spectra; some of the spectral characteristics of this PL were observed previously [4]. We have only found it associated with surfaces and nano-structured materials, particularly surfaces created by fractures [5]. The surfaces from which we observe this PL have a high density of defects. At such high defect densities, interactions between the defects likely become important. We suggest that Förster Resonance Energy Transfer (FRET) [6, 7] and other energy transfer mechanisms lead to the broad distribution in lifetimes observed. In particular, energy transfer between randomly distributed donors (also called sensitizers) and acceptors leads to stretched exponential decay models [8-10]. We develop a way to fit such models, accounting for instrument response and low photon count histograms. We find that the model fits our data well, suggesting that FRET may play a role in the novel photoluminescence in laser damage-prone fused silica.

THEORY AND EXPERIMENT

Stretched exponential decay model

The high defect densities required to explain nanosecond-scale laser damage lead to small average spacing between defects. For the value mentioned above of $n = 10^{19}$ cm^{-3}, the average distance is just under 5 nm. For higher densities obtained for smaller absorption cross-sections, distances between defects will be smaller. At these distances, energy transfer via dipole-dipole interactions may affect the temporal characteristics of the fluorescence lifetime decays.

The expected fluorescence lifetime decay of a set of donor defects in the presence of a randomly distributed acceptor species has been extensively studied. The model was first worked out in the three-dimensional case by Foerster [9]. Additional cases have been worked out subsequently [8]. The generalization to fractal dimensions was found in Klafter et al. [10]. In each case, a stretched exponential model was obtained.

Consider an energy transfer transfer rate between a donor and acceptor of the form

$$\omega(R) = \frac{1}{\tau}\left(\frac{R_0}{R}\right)^s \tag{0.1}$$

The rate of emission from a donor species in the presence of an acceptor species with concentration n is [8]:

$$k(t) = \exp\left[-V_\Delta n R_0^\Delta \Gamma(1-\Delta/s)(t/\tau)^{\Delta/s}\right] \tag{0.2}$$

V_Δ is the volume of unit sphere in Δ dimensions. This expression may also be used for fractal dimensions Δ [10].

By combining many of these parameters, a simplified fitting model may be used:

$$P(t) = A\exp\left[-(t/\tau)^\beta\right] \tag{0.3}$$

For using this model to fit experimental data obtained via time-correlated single photon counting (TCSPC) [11-13], it is necessary for the model to account for the finite bin width of the measurements, and the instrument response. Wraparound of the fluorescence decay due to the high repetition rate of the lasers used for TCSPC must also be taken into account. If the duration of each time bin is Δt, and the time to the next excitation pulse is $N\Delta t$, then integrating Eq. (0.3) we obtain for the ith bin

$$P_i = \sum_{j=0}^{\infty} \frac{1}{\Gamma(\beta^{-1})}\left\{\Gamma\left[\beta^{-1}, \left(\frac{(jN+i+1)\Delta t}{\tau}\right)^\beta\right] - \Gamma\left[\beta^{-1}, \left(\frac{(jN+i)\Delta t}{\tau}\right)^\beta\right]\right\} \tag{0.4}$$

$\Gamma(a,x)$ is the incomplete gamma function, and $\Gamma(a)$ is the gamma function. The functional form of (0.4) is then convolved with a measured instrument response (IRF_i), and fitted to the data using a maximum likelihood procedure as described in [12]. One additional fitting parameter in addition to A, τ, and β is a temporal shift in the position of the IRF.

Experiment

The fluorescence lifetime decays of fused silica surface flaws are measured as described previously [3]. Briefly, a 405 nm pulsed laser diode with 150 ps pulse widths (LDH-405, Picoquant, Berlin, Germany) was focused using a high numerical aperture objective onto a 2N Vicker's indentation. The fluorescence was collected using the same objective, filtered (using a 580 nm bandpass filter), and measured using a single photon counting avalanche photodiode (PDM module, Micro Photon Devices, Bolzano, Italy). The time difference between the laser pulses and the fluorescence photons is measured using a PicoHarp 300 (Picoquant, Berlin, Germany). The instrument response is measured by exciting Silver nanoparticles dried on a fused silica substrate.

RESULTS

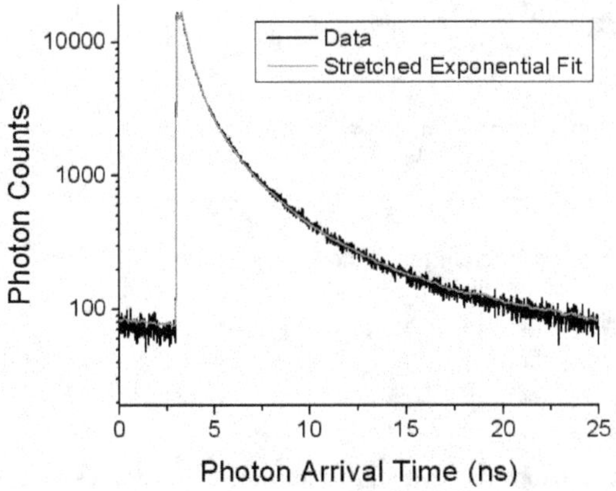

Figure 1: Fluorescence lifetime decay of 2N silica Vicker's indentation is fitted to stretched exponential fit, accounting for instrument response and finite bin widths.

In Figure 1, we show the fluorescence lifetime curve measured from a 2N Vicker's indentation on fused silica. The stretched exponential model developed in the previous section fits well to the fluorescence lifetime decay. The quality of fit obtained with the four parameters of the stretched exponential fit (A, τ, β, and IRF shift; χ^2=1.6) is comparable to that obtained by a four-exponential fit with varying lifetimes (10 parameters; χ^2=1.2) and a four-exponential fit with fixed lifetimes of 0.04, 0.2, 1.0, and 5 ns (6 parameters; χ^2=2.0). The normalized residuals calculated from the stretched exponential fit are similar to those of the other fits (Figure 2). The fit is good over two orders of magnitude in photon arrival time. The high quality fit of the stretched exponential with fewer parameters is consistent with the model described above where energy transfer interactions play an important role in the temporal characteristics of the fluorescence lifetime decay of the novel photoluminescence or fluorescence from fused silica surface flaws.

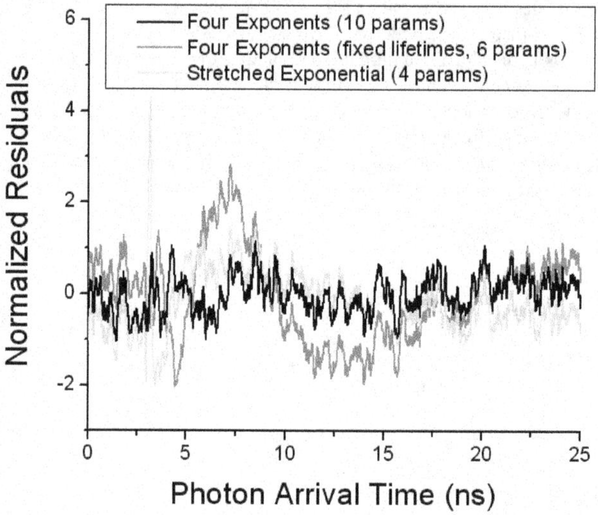

Figure 2: Normalized residuals plotted for three different fits of the data in Figure 1. The fits include a four-exponential fit with 10 parameters, a four-exponential fit with fixed lifetimes and 6 parameters, and the stretched exponential fit with 4 parameters.

DISCUSSION AND CONCLUSION

We have proposed that the distributed fluorescence lifetimes measured in fused silica surface flaws are related energy transfer interactions within a high density of defects. The stretched exponential model naturally results from these models, and is shown to fit the data well with fewer parameters than other common decay models. These results offer preliminary support for this model. Further experiments detailing spectral characteristics and varying concentrations of defects will shed further light on the role of energy transfer interactions.

ACKNOWLEDGMENTS

This work performed under the auspices of the U.S. Department of Energy by Lawrence Livermore National Laboratory under Contract DE-AC52-07NA27344. We thank Rusty Steele for assistance in the preparation of the fused silica samples.

REFERENCES

1. Papernov, S. and A.W. Schmid. *Laser-induced surface damage of optical materials: absorption sources, initiation, growth, and mitigation.* in *Laser-Induced Damage in Optical Materials: 2008*. 2008. Boulder, CO, USA: SPIE.
2. Carr, C.W., J.D. Bude, and P. DeMange, *Laser-supported solid-state absorption fronts in silica.* Physical Review B (Condensed Matter and Materials Physics), 2010. **82**(18): p. 184304.
3. Laurence, T.A., J.D. Bude, N. Shen, T. Feldman, P.E. Miller, W.A. Steele, and T. Suratwala, *Metallic-like photoluminescence and absorption in fused silica surface flaws.* Applied Physics Letters, 2009. **94**(15): p. 151114.
4. Kucheyev, S.O. and S.G. Demos, *Optical defects produced in fused silica during laser-induced breakdown.* Applied Physics Letters, 2003. **82**(19): p. 3230-3232.
5. Miller, P.E., J.D. Bude, T.I. Suratwala, N. Shen, T.A. Laurence, W.A. Steele, J. Menapace, M.D. Feit, and L.L. Wong, *Fracture-induced subbandgap absorption as a precursor to optical damage on fused silica surfaces.* Opt. Lett., 2010. **35**(16): p. 2702.
6. Dexter, D.L., *A Theory of Sensitized Luminescence in Solids.* Journal of Chemical Physics, 1953. **21**(5): p. 836-850.
7. Forster, T., *Energiewanderung und Fluoreszenz.* Die Naturwissenschaften, 1946. **6**: p. 166-175.
8. Blumen, A., *Excitation Transfer From A Donor To Acceptors In Condensed Media - A Unified Approach.* Nuovo Cimento Della Societa Italiana Di Fisica B-General Physics Relativity Astronomy And Mathematical Physics And Methods, 1981. **63**(1): p. 50-58.
9. Forster, T., *Experimentelle Und Theoretische Untersuchung Des Zwischenmolekularen Ubergangs Von Elektronenanregungsenergie.* Zeitschrift Fur Naturforschung Section A-A Journal Of Physical Sciences, 1949. **4**(5): p. 321-327.
10. Klafter, J. and A. Blumen, *Fractal Behavior In Trapping And Reaction.* Journal Of Chemical Physics, 1984. **80**(2): p. 875-877.
11. Becker, W., *Advanced Time-Correlated Single Photon Counting Techniques.* Springer Series in Chemical Physics, ed. J.P.T.a.W.Z. A. W. Castleman. Vol. 81. 2005, Heidelberg, New York: Springer.
12. Laurence, T.A. and B.A. Chromy, *Efficient maximum likelihood estimator fitting of histograms.* Nature Methods, 2010. **7**(5): p. 338-339.
13. Zander, C., K.H. Drexhage, K.T. Han, J. Wolfrum, and M. Sauer, *Single-molecule counting and identification in a microcapillary.* Chemical Physics Letters, 1998. **286**(5-6): p. 457-65.

Laser Ablation and Deposition, and Surface Modification

Mater. Res. Soc. Symp. Proc. Vol. 1365 © 2011 Materials Research Society
DOI: 10.1557/opl.2011.1385

Metal and Metal-oxide Nanoparticle Synthesis by Laser Ablation of Aqueous Aerosols

Kristofer L. Gleason[1], John W. Keto[2], Desiderio Kovar[3], and Michael F. Becker[1]

[1] Department of Electrical and Computer Engineering, University of Texas at Austin, Austin, TX, U.S.A.

[2] Department of Physics, University of Texas at Austin, Austin, TX, U.S.A.

[3] Department of Mechanical Engineering, University of Texas at Austin, Austin, TX, U.S.A.

ABSTRACT

We present a scalable, continuous manufacturing method of nanoparticle production based on laser ablation of an aerosol generated from an aqueous precursor solution. A Collison nebulizer is used to generate a mist of ~10 μm diameter water droplets containing dissolved transition metal salts, suspended in 1 atmosphere of buffer gas. Water from the droplets quickly evaporates, leaving solid particles ~2 μm in diameter for a typical solution concentration. These microparticles are then ablated by a pulsed KrF excimer laser (10 ns, $\lambda = 248$ nm, 2 J/cm^2 at focus). Ablation results in plasma breakdown of the microparticle and photothermal decomposition of the precursor material. Following ablation, nanoparticles 5-20 nm in diameter are formed and collected. For $AgNO_3$ ablated in He gas, metal Ag nanoparticles were produced. For $Cu(NO_3)_2$ ablated in He, crystalline Cu_2O nanoparticles were produced. For $Ni(NO_3)_2$ ablated in He, crystalline NiO nanoparticles were produced. A combination of $AgNO_3$ and $Cu(NO_3)_2$ ablated in a reducing atmosphere of 10% H_2 and 90% He yielded Ag-Cu alloy nanoparticles. In contrast to conventional wet-chemical synthesis processes, our nanoparticles are formed 'bare,' without surfactants or organic material contaminating the surface. Owing to their small size and high free surface area, nanoparticles produced by this process are ideally suited for applications that include catalysis and facilitated transport membranes.

INTRODUCTION

Nanoparticle synthesis is routinely accomplished by wet chemical techniques [1]. However, these protocols require use of a surfactant or polymer capping agent to stabilize particles from agglomeration. For applications requiring a clean particle surface, more elaborate synthesis routes are necessary. For example, spray pyrolysis [2] produces particles by thermal decomposition of an aqueous aerosol of precursor compounds flowing through a tube furnace. A residence time of several seconds is usually required for complete reaction. During this time, particles are subject to Brownian coagulation, which limits the minimum particle size. In the laser ablation of microparticle aerosol (LAMA) technique developed by our lab [3], nanoparticles are formed quickly via a laser driven shock process, and much shorter residence times are possible. We routinely produce Ag nanoparticles with 7 nm mean diameters, and a narrow size distribution. However, the technique requires feedstock material in the form of a micron-sized powder aerosol, adding substantially to the cost of production.

In the present work, we introduce a new laser ablation technique that utilizes aqueous aerosol feedstocks rather than powders. Results of transmission electron microscope (TEM), energy dispersive spectroscopy (EDS), and electron diffraction observations are presented for several experiments involving transition metal nitrate salts as precursors. Results involving pure nitrate salts ablated in an inert atmosphere are presented first. Next, experiments involving a mixture of Ag/Cu nitrate salts ablated in both inert and reducing atmospheres are presented. The results presented here are preliminary in nature, and are intended as a proof of concept. Additional work is required in order to optimize the process for large scale nanoparticle production.

EXPERIMENTAL DETAILS

A detailed description of the laser ablation system appears elsewhere [3]; a brief summary is presented here. A schematic of the apparatus is shown in figure 1. A Collison nebulizer [4] was used for the aerosol feed. The aerosol droplets had diameters on the order of 10 μm. The droplets quickly lose water to evaporation in route to the ablation cell, leaving a salt particle roughly 2 μm in diameter to be ablated by the laser. Nitrate salts (ACS reagent grade, Sigma-Aldrich) of Ag, Ni, Cu, and 50% Ag/Cu (mol/mol) were prepared in 1.2 M solution with deionized water. He gas (99.9% purity) was supplied by Linde; H_2 (99.5% purity) was supplied by Matheson Tri-Gas. He and 10% H_2/He gasses were supplied to the nebulizer at 3000 sccm. The laminar flow ablation cell is designed to handle a flow of only 200 sccm. To accommodate this requirement, the nebulizer output flow was split, with the flow in excess of 200 sccm discarded. A KrF excimer laser (Lumonics PM-848, 248 nm wavelength, 12 ns pulse width, 200 Hz repetition rate) was used to ablate the feedstock aerosol. A pair of cylindrical lenses with long focal lengths were used to define an ablation region 1.5 mm wide, 4 mm tall, and 5 mm deep. The mean fluence within the ablation region was 2 J/cm^2; everywhere within the ablation region, the fluence exceeded 1 J/cm^2. The aerosol feed was directed through the ablation region using a 1 by 4 mm nozzle. A laminar sheath of clean gas surrounded the aerosol feed. For the experiments described here, the feed aerosol density was intentionally kept very low, estimated to be within 10^4 to 10^5 particles/cm^3; we estimate less than 0.2% extinction of the incident laser energy by the aerosol. After ablation, particles were collected for analysis using a supersonic impactor [5]. A 300 μm diameter flat-plate nozzle was used, with a stagnation pressure of 760 Torr. Downstream pressure was maintained by a mechanical pump and roots blower at 280-300 mTorr. Particles were impacted onto Cu or Au TEM grids coated with lacey carbon, attached to a flat-plate stage located 2.5 mm from the nozzle.

RESULTS AND DISCUSSION

Single component precursors: Ag, Cu, Ni

TEM micrographs of typical particles produced from $AgNO_3$, $Cu(NO_3)_2$, and $Ni(NO_3)_2$ feedstocks ablated in He gas appear in figure 2, with corresponding EDS spectra in figure 3. From the $AgNO_3$ feedstock, spherical particles in the 5-20 nm range are formed. The EDS spectrum clearly indicates Ag, and does not indicate N or O. Furthermore, analysis of the Fourier power spectra for these micrographs indicates a face centered cubic (FCC) structure with $a = 4.1$ Å, consistent with metallic Ag. Therefore, it appears the $AgNO_3$ precursor decomposed completely upon ablation, yielding Ag nanoparticles.

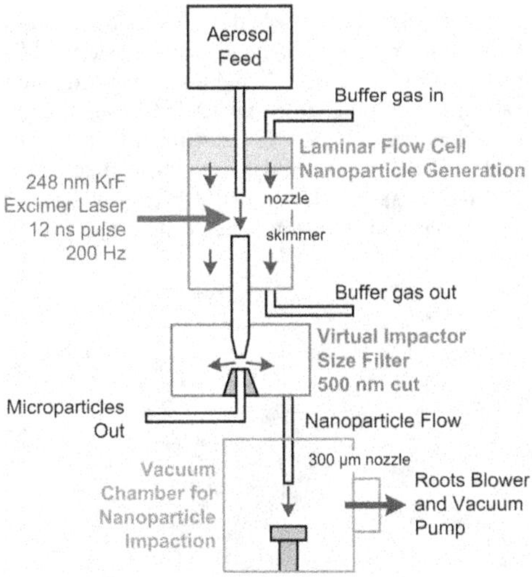

Figure 1. Schematic of laser ablation system for nanoparticle production.

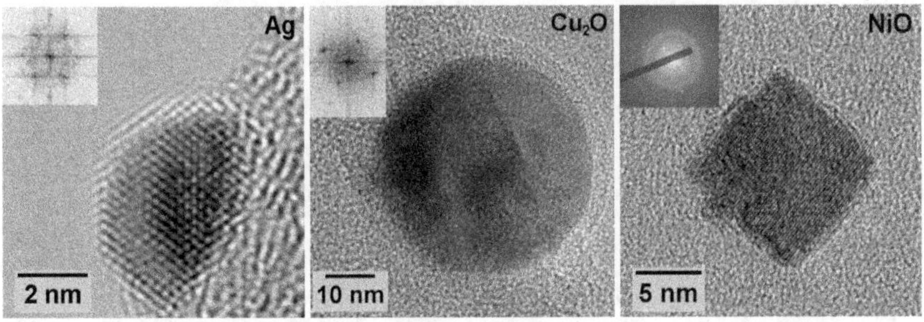

Figure 2. TEM micrographs of typical Ag, Cu_2O, and NiO particles produced by ablating $AgNO_3$, $Cu(NO_3)_2$, and $Ni(NO_3)_2$ aerosols in He buffer gas. Inset: Fourier power spectra (Ag and Cu_2O) and electron diffraction pattern (NiO).

The $Cu(NO_3)_2$ and $Ni(NO_3)_2$ experiments, on the other hand, do not yield metallic particles. EDS spectra for both show a significant O peak; neither sample shows N. Electron diffraction observations were conducted for both materials. For the $Cu(NO_3)_2$ ablation experiment, diffraction patterns are consistent with the cuprite structure, Cu_2O, with $a = 4.3$ Å. From the micrographs, we can also see an amorphous shell surrounding the crystalline core. We have not positively identified the composition of the shell, but we speculate that it could be CuO formed by oxidation of the particle surface in air [6]. The Cu_2O particles are spherical, typically 10-40 nm in diameter. For $Ni(NO_3)_2$ ablation, diffraction pattern indexing indicates a rock salt structure with $a = 4.2$ Å, consistent with NiO. These particles appear cubic, 5-20 nm on a side, with jagged faces. Results for both experiments indicate decomposition of the precursor salts to form oxide particles. Formation of oxide particles could result from either of two different processes: oxidation in the ablation cell at the time of particle formation, or oxidation after particles are exposed to the ambient atmosphere. It is shown in the following section that Cu_2O is likely formed within the ablation cell. On the other hand, the cubic shape of NiO particles suggests that oxidation occurred after particle formation. If oxidation had occurred prior to particle formation, we would expect to see oxide particles with a spherical shape as a result of rapid quenching.

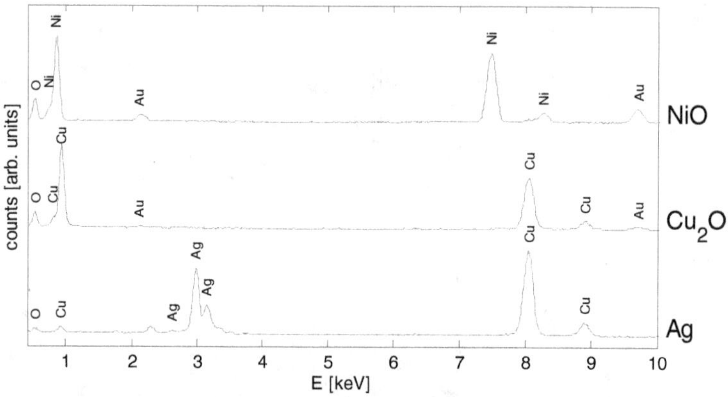

Figure 3. Typical EDS spectra for Ag, Cu_2O, and NiO particles (offset for clarity). Cu TEM grids were used for the Ag experiment, and Au grids were used for the others.

Ag/Cu precursor mixture

In order to investigate the possibility of producing alloy metal particles, we conducted a second set of experiments using 50% molar mixtures of $AgNO_3$ and $Cu(NO_3)_2$. TEM micrographs and EDS spectra for these experiments appear in figures 4 and 5. Our initial experiments were conducted in an inert He atmosphere as before. Particles produced under these conditions were of non-uniform composition and irregularly shaped, with 10-40 nm diameters. There is also evidence of an amorphous layer surrounding the particles. Analyses of Fourier power spectra show *d*-spacings corresponding to metallic Ag and to Cu_2O. Individual particles appear polycrystalline, composed of islands of these two crystal phases. EDS spectra indicate the presence of Ag, Cu, and O.

Ablating the same feedstock in a reducing 10% H_2/He atmosphere produced particles with a very different morphology. These particles are spherical, with diameters from 5-20 nm, and much more uniform than those produced in an inert atmosphere. Electron diffraction experiments gave results consistent with FCC crystal structure, and *d*-spacings related to Cu_2O were no longer found. A thin amorphous layer appears on the particle surface. O is found in the

EDS spectra in addition to Ag and Cu, although at lower intensities compared with particles produced in an inert atmosphere. We speculate that the amorphous layer is an oxide phase.

The difference in particle morphology between these experiments gives some clues about how the particles form. Based on our observations, we believe the $Cu(NO_3)_2$ decomposes partially to form Cu_2O in an inert atmosphere, but decomposes to metallic Cu in a reducing atmosphere. The irregularly shaped particles formed in the inert atmosphere are consistent with rapid condensation of Cu_2O along with Ag metal. The reducing atmosphere, on the other hand, yields metallic Cu which can form a liquid solution with Ag during the initial condensation, ultimately forming spherical particles.

The Ag-Cu system is eutectic, with a miscibility gap spanning nearly the entire range of compositions up to the melting temperature [7]. Ag-Cu solutions phase segregate as they solidify. However, metastable non-equilibrium mixtures can be formed by rapid quenching [8], which is the case for our process. Additional work is in progress to further characterize the phase composition of particles produced by this process.

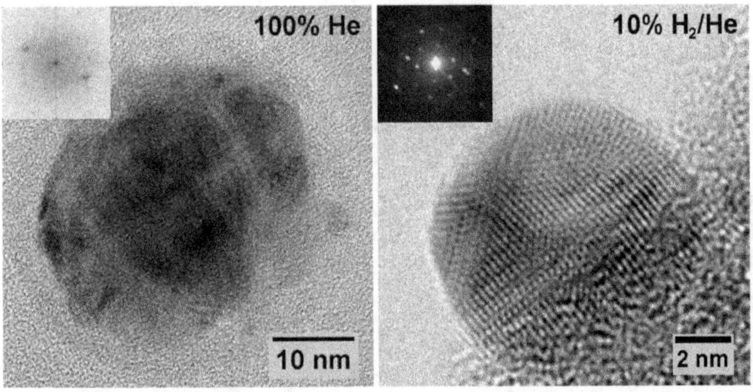

Figure 4. TEM micrographs of typical particles produced by ablating 50% $AgNO_3/Cu(NO_3)_2$ (mol/mol) in 100% He and in 10% H_2/He atmospheres. Inset: Fourier power spectrum (100% He) and electron nanobeam diffraction pattern (10% H_2/He). The nanobeam diffraction pattern corresponds to a different particle than shown in the micrograph.

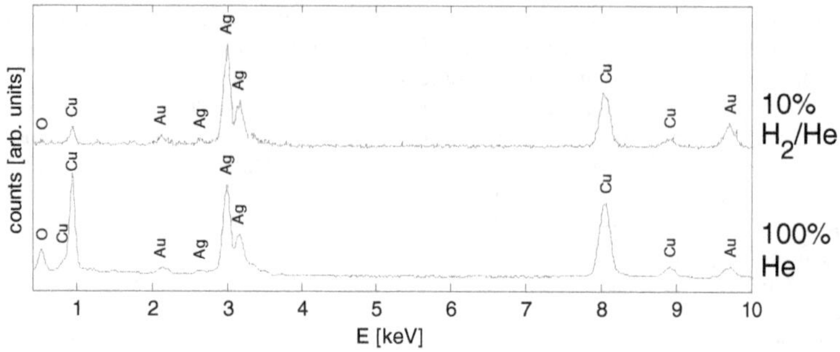

Figure 5. Typical EDS spectra for particles produced by ablating 50% $AgNO_3/Cu(NO_3)_2$ (mol/mol) in 100% He and in 10% H_2/He atmospheres. Au TEM grids were used for both experiments.

CONCLUSIONS

We have presented a new laser ablation technique for synthesizing nanoparticles from aqueous aerosol feedstocks that are free from any organic surfactant coating. It was shown that $AgNO_3$ decomposes upon ablation to form metal Ag nanoparticles, while $Cu(NO_3)_2$ and $Ni(NO_3)_2$ decompose to form oxide nanoparticles. A mixed $AgNO_3/Cu(NO_3)_2$ system was investigated. In an inert atmosphere, irregular composite particles containing Ag and Cu_2O were produced. In a reducing atmosphere, $Cu(NO_3)_2$ decomposed to metallic Cu which mixed with Ag during nucleation, forming Ag-Cu alloy particles.

ACKNOWLEDGEMENTS

This work was supported by the National Science Foundation NIRT Award No. CBET 0708779.

REFERENCES

1. P.Y. Silvert, R. Herrera-Urbina, N. Duvauchelle, V. Vijayakrishnan and K.T. Elhsissen, J. Mater. Chem. **6**, 573 (1996).
2. D. Majumdar, T.T. Kodas and H.D. Glicksman, Adv. Mater. **8**, 1020 (1996).
3. W. Nichols, J. Keto, D. Henneke, J. Brock, G. Malyavanatham, M. Becker and H. Glicksman, Appl. Phys. Lett. **78**, 1128 (2001).
4. K.R. May, J. Aerosol Sci. **4**, 235 (1973).
5. J. Fernández de la Mora, S.V. Hering, N. Rao and P.H. McMurry, J. Aerosol Sci. **21**, 169 (1990).
6. R. Woods, B.D. Fleming, A.N. Buckley and B. Gong, ECS Trans. **28** (6), 51 (2010).
7. P. Subramanian and J. Perepezko, J. Phase Equilib. **14**, 62 (1993).
8. O. Pandey, N. Mishra, C. Ramachandra, S. Lele and S. Ojha, Metall. Mater. Trans. A **25**, 2518 (1994).

Mater. Res. Soc. Symp. Proc. Vol. 1365 © 2011 Materials Research Society
DOI: 10.1557/opl.2011.1222

Formation of Foamy Coatings by Laser Ablation of Glass-ceramic Substrates in the Nanosecond Regime Substrate - Temperature and Wavelength Dependence

Daniel Sola[1], Andrés Escartín[2], Jose I. Peña[2]

[1]Centro de Física de Materiales, Universidad del País Vasco-CSIC
Departamento de Física Aplicada I
Alda. de Urquijo s/n
48.013 Bilbao, Spain
[2]Instituto de Ciencia de Materiales de Aragón, Universidad de Zaragoza-CSIC
Departamento de Ciencia y Tecnología de Materiales y Fluidos
C/ María de Luna, 3
50.018 Zaragoza, Spain

ABSTRACT

A study of the formation mechanisms of foamy coatings on the surface of glass-ceramic substrates produced by laser ablation is presented. Three laser systems emitting at 1064, 532 and 355 nm with pulse-widths in the nanosecond range were used. In the NIR range the formation of the coating is only possible when the temperature of the surface is higher than 300 °C. In this case, the generation is related to an increase of the layer in liquid-phase produced in the interaction zone. However, when the sample is machined at 532 or 355 nm, it is not necessary to heat the whole surface to be processed. In this case, the local temperature and the pressure exerted over the interaction zone produce the generation of this coating, obtaining the layer at room temperature. Furthermore, the coating can be produced at higher speeds. In this way, it is possible to reduce the energetic cost improving the efficiency of the process.

Morphology, microstructure, composition and thermal properties of the layer are described.

INTRODUCTION

Laser processing is of great interest in the field of optics, electronics, microelectronics, aerospace and medicine. This technique is cost-effective compared to traditional methods and it may be applied to a wide range of substrates, such as metals, ceramics and semiconductors [1]. In the field of materials processing by laser several methods such as laser machining, micromachining, marking, drilling and pulsed laser deposition, have been developed in the last two decades [2,3].

The appearance of techniques for generating short and ultrashort laser pulses, ranging from tens of nanoseconds to a few femtoseconds, has allowed the availability of more powerful systems, with power densities that can reach TW/cm^2. These laser systems, with better features and lower prices, offer a high-speed/high-quality tool for laser machining, which is of great interest in both basic and applied research for scientific and technological purposes [3,4].

However, the foundations of the mechanisms involved in laser ablation are far from being well established. It is known that laser ablation depends on laser wavelength, optical features of laser beam, pulse-width regime and optical-thermal-mechanical properties of the substrate. Some theoretical descriptions have been developed by many authors to generalize the stages of the ablation process: laser radiation absorption, heat transfer to the target, evaporation and gas-dynamic of the vapour [4-14]. Glass and glass ceramic substrates are commonly used in a variety of applications such as lasing systems, opto-informatic devices, microoptical components, mirrors and waveguides, the optical properties of which are of great interest [15-18].

EXPERIMENTAL

Laser Processing

To study and compare the machining process three diode-pumped Q-Switch laser systems were used operating at different wavelengths: 1064, 532, 355 nm. The three laser systems are equipped with a programmable galvanometer at the output of the cavity controlled by CAD software in such a way that the beam can be deflected to perform any predefined pattern. The machining process is controlled by the diodes pump current I_P (in relation to peak power), pulse frequency f, linear speed V and distance between adjacent lines Δ.

As material, a glass-ceramic substrate, Ceran Suprema®, manufactured by Schott was used.

When the sample was processed at 1064 nm, the substrate temperature was modified and controlled from room temperature up to 600°C placing the sample at the focus point of an elliptical mirror with a halogen lamp at the other focus point which allows an increase of the surface temperature above 700°C in a working area of 1cm^2. A thermocouple was used to check the temperature of the substrate. The sample was processed with pulses of 2.45 mJ, irradiance I of 83 GW/cm^2, frequency of 2 kHz, a linear speed of 1 mm/s with a distance between adjacent lines of 10 μm.

When it was machined at 532 and 355 nm, the processing was carried out at room temperature. For 532 nm the processing was performed with pulses of 0.27 mJ, an irradiance of 28 GW/cm^2, using a working frequency of 20 kHz, a linear speed of 25 mm/s and a distance between adjacent lines of 10 μm. For 355 nm the sample was processed with pulses of 130 μJ, an irradiance of 8.5 GW/cm^2, a working frequency of 15 kHz, a linear speed of 60 mm/s and a distance between adjacent lines of 10 μm.

Characterization techniques

The microstructure and composition has been determined by means of scanning electron microscopy (SEM) using a JEOL JSM6400 with EDS analysis. Superficial photography has been carried out with a stereoscope microscope. Diffusivity measurements have been determined between room temperature and 500°C using a Holometrix Thermaflash 2200 laser system. Crystalline phases were determined by XRD analyses with a Bruker D8 Advance diffractometer. Absorbance spectrum and diffuse reflectance were measured using a double beam spectrophotometer UV-Vis-IR Cary 500 Varian.

DISCUSSION

Substrate temperature dependence

When the laser wavelength is on the NIR range, the coating formation depends on the temperature of the substrate [19]. At room temperature, laser machining generates a groove on the surface of 300 μm but with the same laser conditions when the temperature of the surface is higher than 300°C a foamy coating is produced, Fig. 1-(a). The white colour of the coating, produced by the crystallization process carried out during the laser processing, gives a high contrast if compared to the black colour of the former. Furthermore, the surface of the coating is very irregular and the inside is porous with a thickness that can reach about 500 μm, Fig. 1-(b).

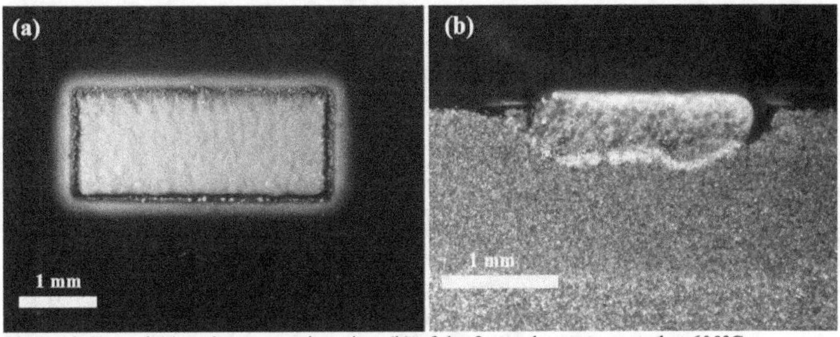

Figure 1. Frontal (a) and cross-section view (b) of the foamy layer generated at 600°C.

During the ablation process, a thin layer of material in liquid-phase is formed in the interaction zone [2], Fig. 2-(a). The recoil pressure produced in the process squeezes the liquid out from the interaction zone and the material is removed from the surface via evaporation and liquid-phase expulsion. The thickness of liquid phase, h_l, and recoil pressure, P_{rec}, can be expressed as [2]

$$h_l \propto (D)^{1/2}\left(\frac{\Delta H_V}{I_a}\right)^{1/4} \qquad (1)$$

$$p_{rec} \approx 10^{-5} I_a \qquad (2)$$

where D is the thermal diffusivity, ΔH_v the enthalpy of vaporization, and I_a the absorbed laser intensity.

Since the variations of thermal diffusivity and enthalpy of vaporization with temperature are negligible, the thickness of the molten layer and the recoil pressure depend mainly on the absorbed irradiance. Taking into account the working conditions and the laser features, in this case the ablation process is a photothermal-mechanical one. Since the melting temperature of the glass ceramic is approximately 1200°C, by increasing the temperature the vibrations of the ions in the lattice increase and the material acquires a pseudo plastic character. This fact produces a diminution in the effective irradiance and so, in the recoil pressure produced, resulting in an increase of the thickness of the layer in liquid-phase present in the interaction zone, Figure 2-(b), generating the foamy coating.

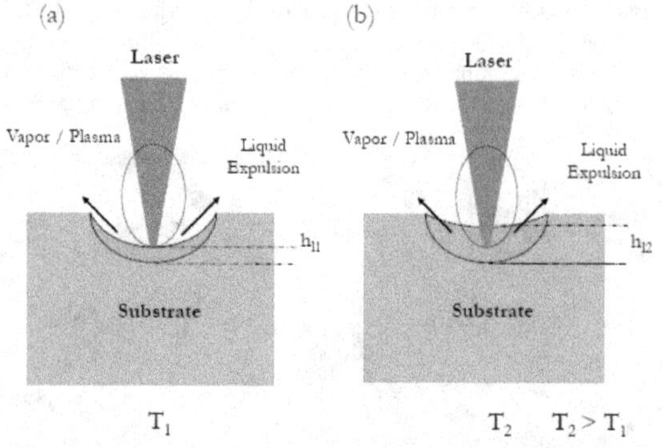

Figure 2. Scheme of ablation process and molten layer of thickness h_l formed in the interaction zone (a). The thickness increases when the surface temperature of the glass-ceramic substrate is heated while is machined using a laser system emitting at 1064 nm.

Wavelength dependence

The formation of the foamy coating is also possible when the wavelength of the laser beam is in the green or in the UV range. Furthermore, it is produced at room temperature, reducing the energetic cost of heating the sample. In this case, the formation of the foamy coating is in relation with the variation of the absorbance of the glass-ceramic substrate with the wavelength. Optical absorption α for 355 nm is around 62 cm^{-1}, 17 times greater than for 1064 nm.

According to the equations (1) and (2), when the substrate processing is carried out with a lower wavelength the pressure increases and a diminution in the thickness of the layer in liquid-phase is produced. Furthermore, the temperature on the surface can be expressed as [6]

$$\overline{T}_n \cong 2T_m(\tau_L f)^{1/2} \tag{3}$$

where $T_m \propto (I_a(D\tau_L)^{1/2})/k$, τ_L is the pulse width, f the working frequency, D the thermal diffusivity and y k the thermal conductivity.

Thus, an increase in the absorbed irradiance produces a higher temperature on the surface of the substrate, $\overline{T}_n$. In this way, laser processing at 532 nm or 355 nm produces a local increase in the temperature of the substrate and in the recoil pressure exerted over the substrate, inducing a growth of the crystalline phase of the glass-ceramic substrate and generating the foamy layer. In order to compare the energetic yield needed in the formation of the coating, the energy delivered per square millimetre was calculated by means of the pulse energy and the dynamic parameters of the process, i.e., frequency, linear speed and distance between adjacent lines resulting in 29.4, 0.86 and 0.42 J/mm^2 for 1064, 532 and 355 nm respectively.

Thus, as the processing is carried out at lower wavelength, the absorption of the laser radiation is favoured, obtaining a foamy coating without the requirement of heating the whole sample and being possible to process at higher speed and entailing a lower energetic cost.

Micro-structural characterization

The elemental composition of the layer was analyzed by semi-quantitative EDS microanalysis, concluding that the composition of the foamy coating is the same of the former. A glass-ceramic substrate consists of a crystalline phase in a vitreous phase. To check possible variations induced in the crystalline phase by the laser processing, a sample of the foamy coating was analyzed by XRD obtaining that there were no modifications in the crystalline phase of the glass-ceramic substrate. Thermal properties of the layer were analyzed comparing the thermal diffusivity of the glass-ceramic with respect to the processed sample. As Fig. 3 shows, the diffusivity was tested between room temperature and 500 °C. The presence of the crystalline phase favours the heat conduction and furthermore, predominates over the pores present in the layer. In this way, the diffusivity of the processed sample is higher than the glass-ceramic one in the whole range of temperatures.

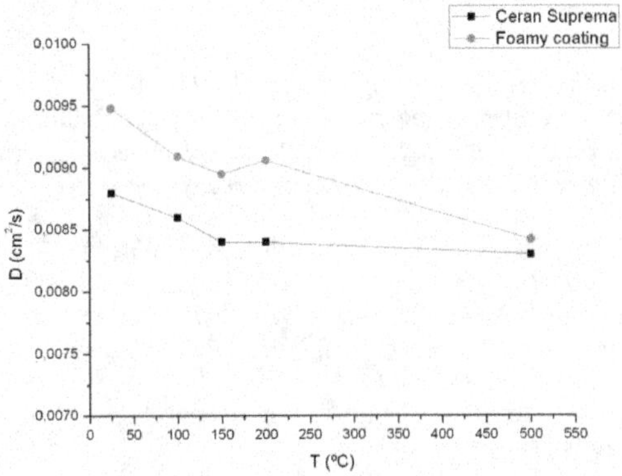

Figure 3. Thermal diffusivity of the foamy coating compared with the glass-ceramic former.

CONCLUSIONS

The formation mechanism of foamy coating as a function of the wavelength of the laser beam has been investigated. It depends on whether the sample is processed with laser beams emitting at 1064 nm or at 532 and 355 nm. In the NIR range the formation takes place only if the surface of the sample is heated beyond 300° C. In this case, a decrease in the recoil pressure leads to an increase in the thickness of the layer in liquid-phase present in the interaction zone generating the foamy coating.

When the sample is processed at 532 or 355 nm, the formation is related to the increase in the absorbance of the glass-ceramic substrate at these wavelengths. In these cases, the better absorption produces a local increase in the recoil pressure and in the temperature of the sample, inducing a growth of the crystalline phase of the glass-ceramic substrate and generating the foamy layer. Furthermore, this process is carried out at room temperature, avoiding the heating of the whole sample and thus, reducing the energetic cost. Furthermore, the coating can be generated at a higher linear speed. The efficiency of the process is also improved by machining with lower wavelengths.

XRD analysis showed no modifications in the crystalline phase of the glass-ceramic substrate. Besides, despite the presence of pores inside the layer, thermal diffusivity test showed that the crystalline phase favours the heat conduction between room temperature and 500 °C.

ACKNOWLEDGMENTS

Daniel Sola thanks the JAE-DOC program, the University of Zaragoza, the BSH Home Appliances Group, the Science and Technology Inter-Ministry commission of Spain and FEDER founds of the EC under project MAT2009-13979-C03-03 for the financial support of his contract.

REFERENCES

1. W. Steen. *Laser material processing*. Springer Verlag (1999).
2. Dieter Bäuerle. *Laser processing and chemistry*. Springer (2000).
3. M. Allmen, A. Blatter. *Laser-beam interactions with materials*. Springer (1995).
4. Y. Pauleau. *Materials surface processing by directed energy techniques*. Elsevier (2006).
5. H. Misawa, S. Juodkazis. *3D Laser microfabrication*. Wiley (2006).
6. E.G. Gamaly, A.V. Rode, B. Luther-Davies. *J. Appl. Phys.* **85** (8) (1999) 4213-4221.
7. A.V. Rode, B. Luther-Davies, E.G. Gamaly. *J. Appl. Phys.* **85** (8) (1999) 4222-4230.
8. N. M. Bulgakova, A.V. Bulgakov. *Appl. Phys. A* **73** (2001).
9. M. A. Anisimov. *Sov. Phys.-Usp.* **17** (1975).
10. B. N. Chichkov, C. Momma, S. Nolte, F. Von Alvensleben, A. Tünnermann. *Appl. Phys. A* **63** (1996).
11. C. Momma, B.N. Chichkov, S. Nolte, F. Alvensleben, A. Tünnermann, H. Welling, B. Wellegehausen. *Optics Communications* **129** (1996).
12. A.V. Bulgakov, N.M. Bulgakova. *Quantum Electronics* **29** (1999).
13. A. Gusarov, I. Smurov. *J. Appl. Phys.* **97** (2005).
14. R.K. Singh, J. Narayan. *Phys. Rev. B* **41** (1990).
15. J.A. Pardo, J.I. Peña, R.I. Merino, R. Cases, A. Larrea, V.M. Orera. *Journal of Non-Crystalline Solids* **298** (2002).
16. R. Balda, R.I. Merino, J.I. Peña, V.M. Orera, J. Fernández. *Optical Materials* **31** (2009).
17. R. Balda, R.I. Merino, J.I. Peña, V.M. Orera, M.A. Arriandiaga, J. Fernández. *Optical Materials* **31** (2009).
18. R. Balda, J. Fernández, I. Iparraguirre, J. Azkargorta, S. García-Revilla, J.I. Peña, R. I. Merino, V.M. Orera. *Optics Express* **17** (2009).
19. D. Sola, J.I. Peña. *Applied Surface Science* **255** (2009) 5322-5328.

Mater. Res. Soc. Symp. Proc. Vol. 1365 © 2011 Materials Research Society
DOI: 10.1557/opl.2011.889

Laser direct writing of graphene patterns

J.B. Park, W. Xiong, Z.Q. Xie, M. Mitchell, Y. Gao, M. Qian, Y.F. Lu

Department of Electrical Engineering, University of Nebraska-Lincoln, Lincoln, NE 68588-0511

ABSTRACT

Rapid growth of single-layer graphene using laser-induced chemical vapor deposition (LCVD) with a visible CW laser ($\lambda = 532$ nm) irradiation at room temperature was investigated. In this study, an optically-pumped solid-state laser with a wavelength of 532 nm irradiates a thin nickel foil to induce a local temperature rise, thereby allowing the direct writing of graphene patterns about ~10 μm in width with high growth rate on precisely controlled positions. It is demonstrated that the fabrication of graphene patterns can be achieved with a single scan for each graphene pattern using LCVD with no annealing or preprocessing of the substrate. The scan speed reaches to about ~50 um/s, which indicates that the graphene pattern with 1:1 aspect ratio (x:y) can be grown in 0.2 sec. The patterned graphene on nickel was transferred to SiO_2/Si substrate for fabrication of electrical circuits and sensor devices.

INTRODUCTION

Graphene is a really promising material in emerging electronic technology due to their outstanding electrical [1], mechanical [2,3], thermal [4], and optical properties [5]. 2-dimentional sheet can be used with current semiconductor manufacturing processes. Graphene will be the core material for flexible and transparent electronics [6], including displays, solar cells, sensors, and electrodes. Currently, graphene can be obtained through several different approaches such as mechanical exfoliation [7,8], annealing of SiC [9,10], and chemical vapor deposition (CVD) [11-14].

CVD is the most promising method due to the benefits of large-scale production of high quality graphene and transfer the graphene to arbitrary substrates for various applications [11]. Especially, large-scale patterns can be achieved by patterning of catalytic metal films before growth [12] or direct patterning of graphene after growth [13]. Although the CVD method can achieve large production volume using a roll-to-roll method [14], the fabrication of graphene patterns by CVD methods is a multi-step process which is time-consuming and costly. Generally, CVD methods consist of several steps including annealing, growth, cooling, and patterning process for fabrication graphene patterns, and the growth process needs several minutes up to a few hours [11-14]. Besides, the patterning process also requires expensive masks and time consuming processes. Therefore, more efforts are needed to make the conventional CVD techniques more suitable for electronic applications.

In this study, we developed a direct writing method for fabrication of graphene patterns by laser-induced chemical vapor deposition (LCVD). Line patterns were fabricated on a nickel foil. Localized laser heating on the nickel foil resulted in a rapid local temperature rise at the laser focal point. The growth rate is three orders of magnitude faster than that of the general

CVD methods. Graphene patterns can be fabricated through the single-step process. The graphene patterns on the nickel foil could be transferred onto a SiO_2/Si substrate.

EXPERIMENT

Figure 1 shows a schematic experiment setup of laser direct writing system for graphene patterns. The growth of graphene patterns was carried out in a stainless steel vacuum chamber. Focused continuous wave (CW) laser beam (optically pumped solid state laser; OPSL, Coherent Co. G5, λ = 532 nm) was used as a heat source to increase the local temperature on the nickel foil in the stainless steel chamber. CH_4 was used as the precursor gas with a supply of H_2; and the pressure of the chamber was kept at 500 torr during experiments. After many trials, 10 sccm CH_4 and 5 sccm H_2 were selected as the optimal flow rates during experiments. The laser beam was then normally irradiated to the nickel foil with an objective lens. The laser beam was defocused on the nickel foil with a spot diameter of 10 μm to reduce the power sensitivity. Laser power was controlled at about ~5 W to induce the local temperature rise for graphene growth. For direct writing of graphene patterns, the substrate was moved precisely by a motorized stage with respect to the laser beam. The scan speed of the motorized stage was around 50 μm/s.

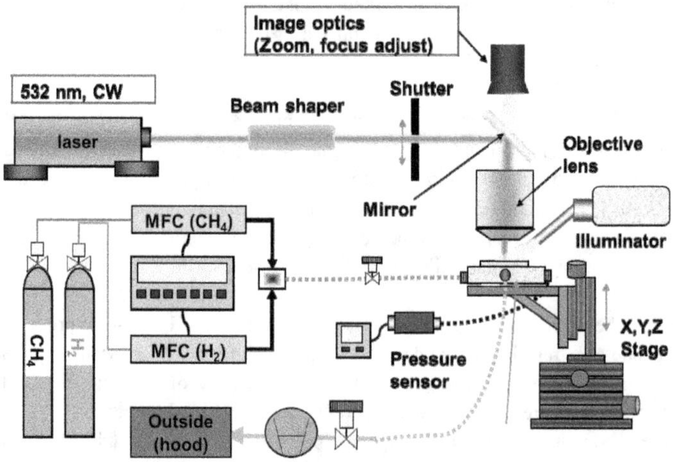

Figure 1. Schematic experiment system for laser direct writing of graphene growth

For transferring graphene to SiO_2/Si substrates (Fig. 2), the graphene-patterned nickel foil was first spin-coated with PMMA at 2500 rpm. The nickel foil was then wet etched using a commercial nickel etching solution for 30-60 min, resulting in a graphene/PMMA film floating on the etchant. Finally, the PMMA layer was dissolved in acetone, leaving the graphene patterns on the SiO_2/Si substrates. This film was then collected manually and rinsed in deionized (DI) water before transferring it onto a SiO_2/Si substrate (with a 300 nm SiO_2 top layer).

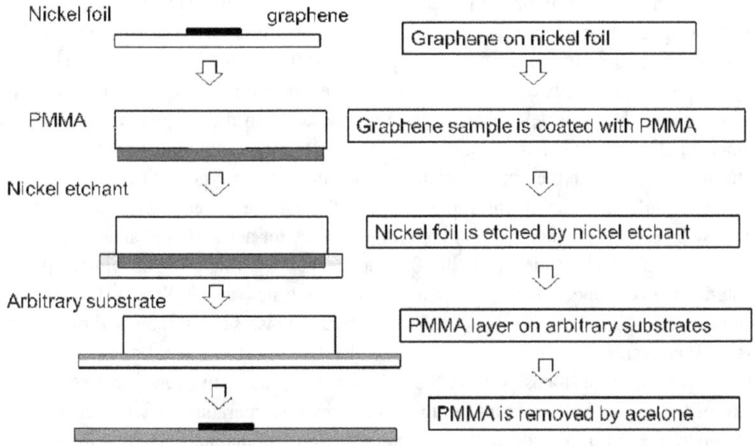

Figure 2. Illustration of graphene transfer procedure from a nickel foil to arbitrary substrates

DISCUSSION

Figure 3(a) shows the optical micrograph of a line pattern of graphene which was fabricated by LCVD on a nickel foil. It is hard to distinguish between the graphene pattern and the Ni substrate because the graphene pattern is highly transparent. The graphene patterns were only observed following the laser scanning path on the illuminated-side of nickel foil, which is different from conventional CVD methods, in which graphene layers are observed on both sides of the foil due to the global heating. Figure 3(b), Magnified SEM image of inside of graphene pattern shows a crumpled graphene sheet. The crumpled graphene is caused by the difference thermal expansion between graphene patterns and the nickel substrate during cooling process [16]. The width difference along the patterns is attributed to a different temperature distribution along the width direction, which results from non-uniform reflectivity of the nickel foil during the laser beam scanning. Raman spectrum in Fig. 3(c) which is measured at 'A' region of Fig. 3(a) clearly shows the evidence of a single-layer graphene. Both the symmetric 2D-band (2691 cm^{-1}) with an FWHM of ~35 cm^{-1} and the high ratio of the peak intensities of 2D- to G-bands (1582 cm^{-1}) (ratio = 3~4) confirm the presence of a mono-layer graphene pattern. The absence of D-band in this spectrum also confirms the high quality of the as-grown graphene. Figure 3(d) show the intensity mapping of 2D-band as shown in Fig. 3(c). The Raman mapping in Figs. 3(d) shows that the intensity of the G-band signals is clearly distinguished from the surrounding non-irradiated areas, thus demonstrating the precisely controlled locality of the graphene pattern in the LCVD process. The Raman intensity distributions of the graphene patterns are relatively uniform in the laser-irradiated areas, also demonstrating the uniformity of the fabricated graphene patterns.

It is noteworthy that the graphene patterns were fabricated with a single scan on the nickel foil without annealing, cooling, or other supportive procedures such as mask patterning. A 10 μm wide line pattern was directly written at a scan speed of ~50 μm/s on the nickel foil. This means that the graphene pattern with a size of 10×10 μm^2 can be rapidly grown in 0.2 sec, which implies that the growth rate is several thousand times faster than that of general CVD methods (up to 2 hours) [11,13]. It is believed that the rapid heating and cooling rates at the local region of the metal foil contribute to the rapid growth of graphene patterns. The fast heating and cooling processes play significant roles in the rapid growth of graphene patterns and can enhance the graphene growth [12,15]. The heating and cooling processes for a 10×10 μm^2 area (a unit area of graphene pattern) have to be finished within 0.2 sec because the line patterns with 10 μm wide are fabricated at the scan speed of 50 μm/s (the cooling rate faster than 5000 °C/s). These heating and cooling processes for a unit area of graphene pattern are also several thousand times shorter than those of conventional thermal CVD methods. Considering inelastic electron-phonon coupling (~ 1ps, [17]), a focused laser beam with high intensity can increase the local temperature of the nickel foil very rapidly up to the growth temperature. After laser irradiation, the local temperature can also drop rapidly through thermal conduction in the entire nickel foil placed in a room temperature environment. Therefore, localized laser heating has much faster heating and cooling rates than those in conventional thermal CVD methods using global heating and cooling process. The single-step fabrication of graphene patterns by LCVD utilizes a rapid heating and cooling mechanism in highly localized regions, which can dramatically reduce the process time.

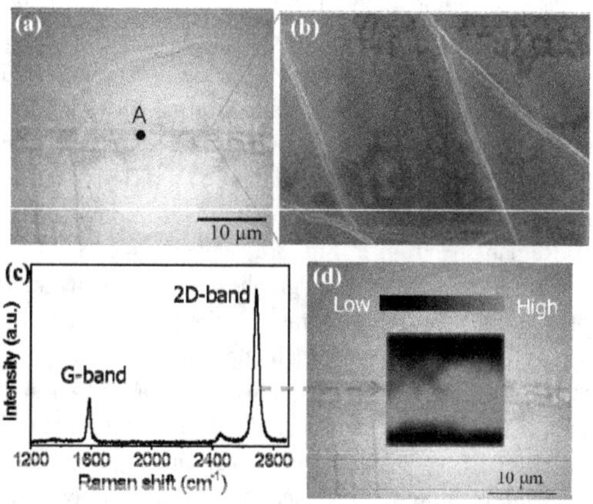

Figure 3. (a) Optical micrograph of a graphene pattern on a nickel foil and (b) Magnified SEM image of crumpled graphene (c) Raman spectrum at 'A' region. (d) Raman intensity mapping of 2D-band (2691 cm^{-1})

Figures 4(a) shows the optical micrographs of a transferred graphene pattern on a SiO_2/Si substrate. The thin top SiO_2 layer under the graphene patterns can enhance the optical sensitivity to human eyes, which allows characterizing the number of graphene layers [18]. The single-layer graphene pattern in Fig. 4(a) is highly transparent and its Raman spectrum in Fig. 4(b) also confirms a single-layer of the as-transferred graphene. It should also be noted that the experimental condition to produce mono-layer graphene patterns is very sensitive to the scan speed. A lower scan speed is more suitable for the fabrication of multi-layer graphene.

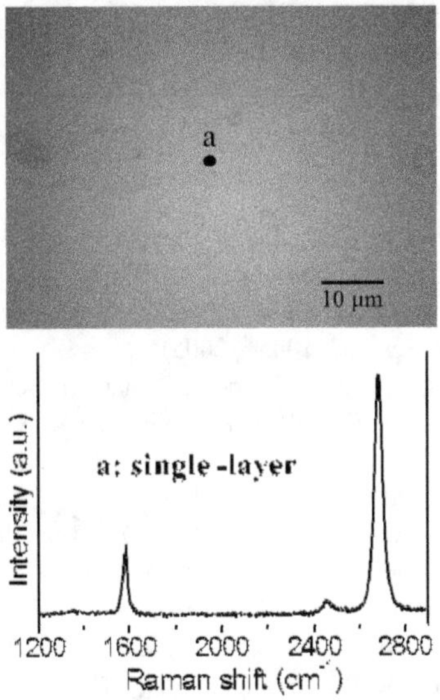

Figure 4. (a) Optical micrograph of transferred single-layer graphene on SiO_2/Si substrate and (b) Raman spectrum of the single-layer graphene at 'a' region.

CONCLUSIONS

Rapid single-step fabrication method was developed for graphene patterning using LCVD. Line patterns of graphene were grown on nickel foils at room temperature with a short process time. No annealing or pattering processes are required for the fabrication of graphene patterns on nickel foils. The significantly enhanced growth rate of graphene is ascribed to the rapid heating and cooling mechanism in the LCVD process. Graphene pattern of 10×10 μm2 area can be rapidly grown in 0.2 sec, which implies that the graphene growth rate is three orders of magnitude faster than the conventional thermal CVD methods. The fabricated line patterns can be transferred to various substrates for different applications, without additional steps removing the graphene from the back-side of nickel foil. It is believed that the LCVD method for graphene patterns will have a significant contribution to the rapid-developing graphene industry.

ACKNOWLEDGMENTS

This research work was financially supported by National Science Foundation (ECCS 0652905 and CMMI 0758199) and Nebraska Research Initiative.

REFERENCES

1. A. K. Geim and K. S. Novoselov, Nature mater. 6, 137 (2007).
2. C. Lee, X. Wei, J. W. Kysar, and J. Hone, Science 321, 385 (2008).
3. S. Bunch, S. S. Verbridge, J. S. Alden, A. M. Zande, J. M. Parpia, H. G. Craighead and P. L. McEuen, Nano Lett. 8, 2458 (2008).
4. R. Prasher, Science 328, 185 (2010).
5. F. Bonaccorso, Z. Sun, T. Hasan and A. C. Ferrari, Nature Photon. 4, 611 (2010).
6. J. K. Wassei and R. B. Kaner, Mater. Today 13, 52 (2010).
7. K. S. Novoselov, D. Jiang, F. Schedin, T. J. Booth, V. V. Khotkevich, S. V. Morozov and Geim, A. K. Proc. Natl. Acad. Sci. 102, 10451 (2005).
8. P. Blake, P. D. Brimicombe, R. R. Nair, T. J. Booth, D. Jiang, F. Schedin, L. A. Ponomarenko, S. V. Morozov, H. F. Gleeson, E. W. Hill, A. K. Geim, and K. S. Novoselov, Nano Lett. 8, 1704 (2008).
9. P. Sutter, Nature mater. 8, 171 (2009).
10. V. Yu. Aristov, G. Urbanik, K. Kummer, D. V. Vyalikh, O. V. Molodtsova, A. B. Preobrajenski, A. A. Zakharov, C. Hess, T. Hänke, B. Büchner, I. Vobornik, J. Fujii, G. Panaccione, Y. A. Ossipyan, and M. Knupfer, Nano Lett. 10, 922 (2010).
11. X. Li, W. Cai, J. An, S. Kim, J. Nah, D. Yang, R. Piner, A. Velamakanni, I. Jung, E. Tutuc, S. K. Banerjee, L. Colombo and R. R. Ruoff, Science 324, 1312 (2009).
12. K. S. Kim, Y. Zhao, H. Jang, S. Y. Lee, J. M. Kim, K. S. Kim, J.-H. Ahn, P. Kim, J.-Y. Choi and B. H. Hong, Nature 457, 706 (2009).
13. Y. Lee, S. Bae, H. Jang, S. Jang, S.-E. Zhu, S. H. Sim, Y. I. Song, B. H. Hong and J.-H. Ahn, Nano Lett. 10, 490 (2010).
14. S. Bae, H. Kim, Y. Lee, X. Xu, J.-S. Park, Y. Zheng, J. Balakrishnan, T. Lei, H. R. Kim, Y. I. Song, Y.-J. Kim, K. S. Kim, B. Özyilmaz, J.-H. Ahn, B. H. Hong and S. Iijima, Nature nanotech. 5, 574 (2010).
15. J. B. Park, S. H, Jong, M. S. Jeong, S. C. Lim and Y. H. Lee, Nanotechnology 20, 185604 (2009).
16. X. Li, Y. Zhu, W. Cai, M. Borsiak, B. Han, D. Chen, R. D. Piner, L. Colombo, and R. S. Ruoff, Nano Lett. 9,4359 (2009).
17. M. A. Al-Nimr, Inter. J Thermophys. 18, 1257 (1997).
18. P. Blakea, E. W. Hill, A. H. Castro Neto, K. S. Novoselov, D. Jiang, R. Yang, T. J. Booth, and A. K. Geim, Appl. Phys. Lett. 91, 063124 (2007).

Mater. Res. Soc. Symp. Proc. Vol. 1365 © 2011 Materials Research Society
DOI: 10.1557/opl.2011.1223

Closed-loop control of a laser assisted carbon nanotube growth process for interconnects in flexible electronics

Yoeri van de Burgt[1,2], Yves Bellouard[1], Rajesh Mandamparambil[2] and Andreas Dietzel[1,2]
[1]Department of Mechanical Engineering, Eindhoven University Technology, Den Dolech 2, Eindhoven, The Netherlands. Tel: +31402472186. E-mail: y.b.v.d.burgt@tue.nl
[2]Holst Centre/TNO – Netherlands Organization for Applied Scientific Research, HTC31, Eindhoven, The Netherlands.

ABSTRACT

A feedback control mechanism based on infrared radiation monitoring coupled with reflectivity information was developed to control the temperature of a laser assisted chemical vapor deposition process for the growth of carbon nanotube forests. An infrared laser operating at 808 nm is focused on a silicon substrate containing a 20 nm-aluminum-oxide layer and a 1.5 nm-iron catalyst layer. The growth takes place in an argon/ hydrogen/ ethylene gaseous environment. SEM and Raman spectroscopy analysis show that good controllability and reproducibility is achieved over multiple experiments.

INTRODUCTION

In recent years, flexible electronics have gained significant importance in the development of flexible displays, wearable devices, intelligent clothes and ambient intelligence [1][2][3]. To mass produce these flexible devices, roll-to-roll manufacturing process involving lamination of various functional multilayer device stacks is used. These functional layers need eventually to be electrically interconnected. Conventional interconnects are made out of metals, pastes or inks filled with metal particles which poses mechanical as well as process related performance issues.

We propose to use carbon nanotubes (CNTs) embedded in a polymer matrix as interconnects for flexible electronics. This is motivated by the excellent electrical properties of CNTs (~10.10^{10} A cm^{-2}) [4] . In this research, a laser assisted chemical vapor deposition method (LACVD) [5] is used to locally grow CNT forests on a silicon substrate in a carbonaceous environment. Laser assisted CVD opens up a range of possibilities for different materials and geometries that are sensitive to heat. CNT length, diameter and density are interesting parameters to tailor for controlling the electrical, mechanical and conformability performances of the interconnects. For example, an obvious requirement is that the CNTs need to be as long as the thickness of the interconnecting layer between the functional layers. The structure, geometry and characteristics of CNTs forests are known to be strongly dependent from the heating profile applied during the growth process [6]. A current limitation of LACVD set-ups [7] [8] is the lack of an accurate temperature control at the laser spot.

Here we present a feedback control mechanism based on infrared radiation monitoring from the CNT growth region triggered by information collected from dynamic reflectivity measurements. The origin of the measured infrared radiation (from 700 to 1700 nm) is essentially of thermal origin and therefore provides indirect information about the average temperature at the spot. In addition to IR radiation, we also monitor additional parameters such as the reflected laser light from the growth site. This information is a direct measure of the nucleation of the CNTs and can be used to apply different temperature strategies before, during and after nucleation. With this technique, we successfully demonstrate that the substrate overheating can be avoided and process reproducibility and controllability is achieved over repetitive experiments.

EXPERIMENTAL

The experimental setup is as shown in Figure 1. The substrates were 4x4 mm silicon substrates with a 20nm aluminum-oxide layer and a 1.5nm iron catalyst layer both deposited by e-beam evaporation. The chamber is a stainless steel cylindrical chamber with a volume of 3.4 L. The chamber has a gas inlet and outlet, a thermocouple placed at the substrate holder, a pressure sensor and an optical window. The laser source is a fiber-pigtailed laser diode (Unique Mode) emitting up to 35W at 808 nm. The laser incidence angle is 35°. The infrared detector that detects the emitted radiation is a GaAs photodiode placed above the substrate. The reflected laser light is measured with a Si photodiode in combination with a laser line band-pass filter.

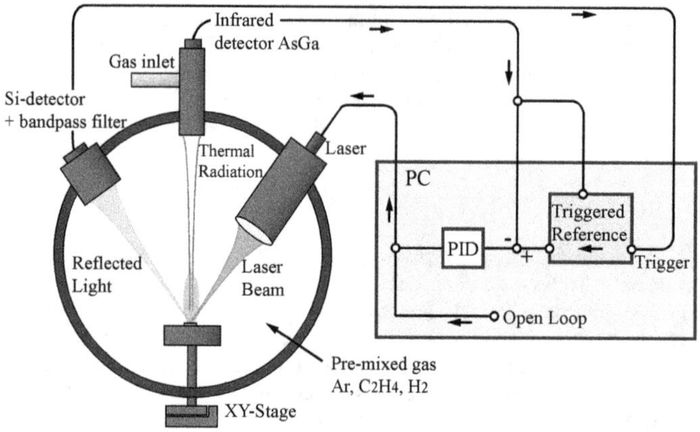

Figure 1. Overview of the laser assisted chemical vapor deposition setup with feedback control. The flow of operation is depicted by arrows. In this particular setup the process is first operated in open loop at which the feedback control is triggered by information from the reflected laser light. The triggered reference set-point is one of the input parameter for the various control schemes.

The communication with the computer is done via a DAQ (NI-DAQ PCI-6221). Simulink is used to design the PID-controller. Three different control strategies and heat profiles are applied. See Table 1. The first strategy (labeled CS1) is to only apply feedback control of the emitted radiation when nucleation has started. This is triggered by a drop in the reflection signal. Before triggering, the process runs open-loop. The second strategy (labeled CS2) also only applies control after nucleation has started but allows for a small increase in emitted radiation. Finally, the third strategy (CS3) is to apply feedback control from the start, so before the nucleation has started. The results are analyzed using a scanning electron microscope (FEI Quanta 600F ESEM) and a Raman spectrometer. (623 nm, Horiba LabRAM HR).

Table 1. Control strategies

Name	**CS 1**	**CS 2**	**CS 3**	**No control**
Description	Triggered control	Triggered control	Overall control	
Experiments	# 1 – 2	# 3 – 6	# 7 - 9	# 10

RESULTS AND DISCUSSION

Control strategy 1 (CS1)

In Figure 2 the detector signals are plotted for experiments #1 and #2 following the control strategy CS1 where control is applied after CNT nucleation. The computer recognizes a drop in reflection indicating a change in surface reflectivity and thus growth of a carbon product. At this time the control is enabled and a reference set-point for the emitted radiation at that particular time is stored. The laser power is then decreased by means of the feedback control of the emitted radiation set-point.

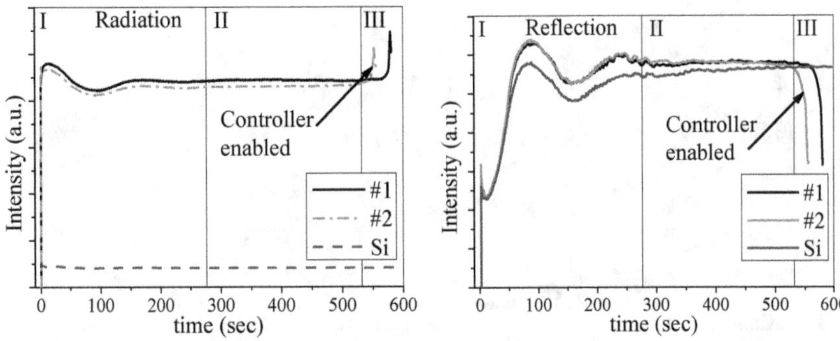

Figure 2. Control strategy CS1: Plot of thermal radiation (left) and reflection of the laser light (right) of experiments #1 and #2. Three regimes can be distinguished. In regime I the emitted radiation and reflection are fluctuating. The fluctuation of this first regime is probably a result of a growing SiO_2 layer. For reference the detector signals on a pure Si substrate are also shown. Regime II is a stable regime and in regime III the drop in reflection and an increase of emitted radiation is visible. The drop in reflection indicates a change in surface reflectivity and thus the growth of a carbonaceous product. The controller is enabled when this drop in reflection is large enough and therefore a small overshoot in emitted radiation is present. The process is stopped when the reflectivity is below a pre-defined value.

We can distinguish three regimes. The first part is where the laser is operated at a constant power of 10 W. In this regime the emitted radiation and reflected laser light are fluctuating. We attribute these signal fluctuations to the growth of a SiO_2 layer, partially transparent at 808 nm that creates interference patterns. Due to the high temperature that is reached, SiO_2 further oxidizes due to remnant presence of oxygen in the chamber (this is a low-vacuum level experiment). The second regime is a stable regime where the radiation and reflection do not change anymore. The temperature of the substrate as a whole is still rising. The third regime starts when the CNT nucleation temperature is reached and this is characterized by a drop in reflection. In the figure this can be found around 540 seconds. A drop in reflection indicates the onset of the CNT growth [5]. At the same time it is observed that the emitted radiation is increasing. This is due to the fact that the substrate is reflecting less laser-light it absorbs more laser energy resulting in an increase in thermal radiation, as detected by the thermal IR detector (see Figure 1 and 2). The controller is enabled at that point and a reference set-point for the radiation is stored. This value is slightly higher than the radiation is just before nucleation because the controller needs some time to detect a significant decrease in the reflection signal. When the controller is enabled it guarantees that no overheating is taking place during growth. The system is automatically stopped when the reflection is below a certain value, in Figure 2 (right) this is half of the maximum value of the reflected laser light. Before nucleation, the two curves (Figure 2) are almost identical apart from a small off-set in the radiation. The nucleation however, starts at a different time for both experiments. This observation shows the importance for a controlled process rather than an open loop, time-based, process. In Figure 3 scanning electron microscopy pictures of two consecutive experiments are shown. These images show a good qualitative reproducibility of the process.

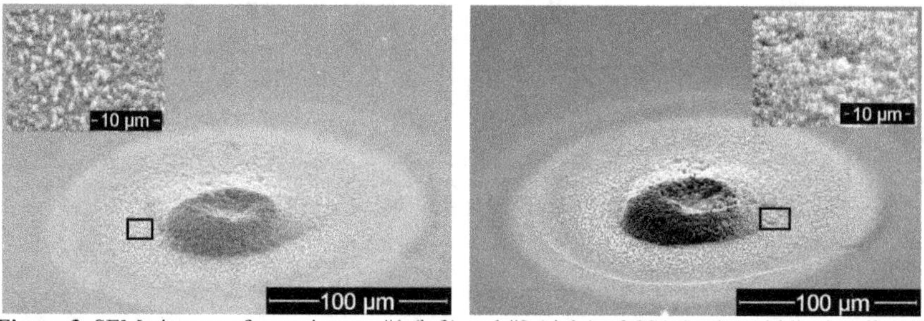

Figure 3. SEM pictures of experiments #1 (left) and #2 (right) of CS1. In the middle the aligned forest of carbon nanotubes is surrounded by a ring structure containing shorter nanotubes. The insets show a zoomed in part of the surrounding nanotubes. Individual as well as aligned carbon nanotubes are clearly visible.

The Raman spectra for this set of experiments are shown in Figure 4 (right). In the range 200 – 1000 cm^{-1} a detailed zoom is shown for both experiments in the inset graph. The two spectra are quite similar which indicates a good reproducibility of the process. We also notice a very large D-band (around 1370 cm^{-1}) that is characteristic for sp^3 hybridized carbon (diamond, amorphous carbon) or defects in the sp^2 hybridized carbon [9]. The nanotubes therefore have defects and are covered with an amorphous carbon layer. This is also verified by the SEM pictures from Figure 3. The G-band is typical for graphite and (multi-walled) carbon nanotubes and is close to 1582 cm^{-1}. The last peak, the G'-band (or D*-band) occurs around 2650 cm^{-1} and is the overtone band of the D-band. This peak corresponds to a double resonance of the D-band and is not depending on the crystal defects [10]. Since sp^3 hybridized carbon does not have a G'-band peak this indirectly confirms the presence of CNTs or CNFs.

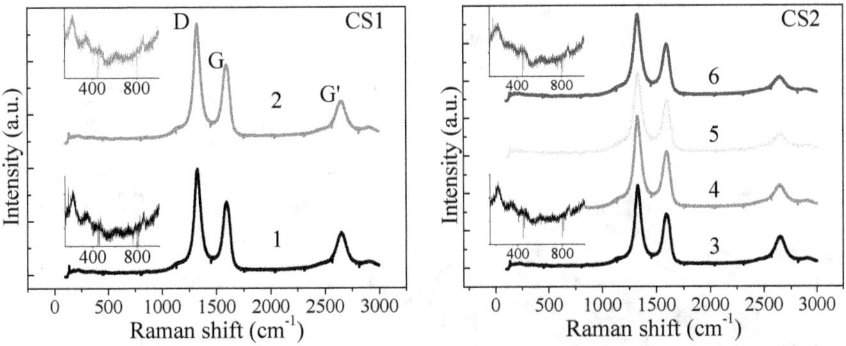

Figure 4. Raman spectra for experiments 1 and 2 of CS1 (left) and 3 – 6 of CS2 (right) with the laser focused at the top part. The characteristic D and G bands as well as the G' are clearly visible. The insets show a zoomed part of the 200 – 1000 cm^{-1} range.

Control strategy (CS2)

In CS2, the experiments (# 3-6) were conducted allowing the radiation to increase right after nucleation. This was done by setting the radiation set-point at twice the value that it had before nucleation. As a result the amount of added energy to the process was higher. However, substantial overheating should still be avoided. Two SEM pictures are shown in Figure 5 corresponding to experiments #3 and #4 of CS2. The Raman spectra corresponding to this set of experiments are shown in Figure 4 (right). From these figures it is clear that aligned CNTs can be reproducibly grown.

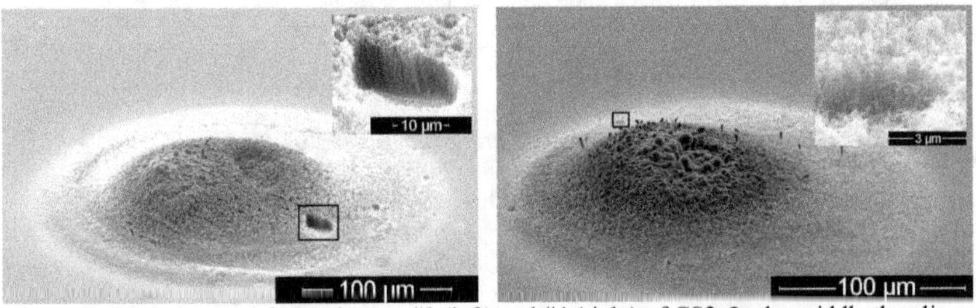

Figure 5. SEM pictures of experiments #3 (left) and #4 (right) of CS2. In the middle the aligned forest of carbon nanotubes is surrounded by a ring structure containing shorter nanotubes. The insets show a zoomed in part of the surrounding nanotubes and aligned nanotubes are shown.

Control strategy (CS3)

The final set of experiments (# 7 - 9) was performed with the radiation feedback control acting on the process from the start of laser exposure. The radiation emitted from the substrate at 4 seconds was measured and used as the set-point value. This control strategy ensured that when the nucleation starts, the controller immediately acts on the process allowing no overshoot to occur. The Raman spectra for this set of experiments are shown in Figure 6 (left). In this set also an uncontrolled experiment (#10) is added for comparison. The D/G band value is usually a measure for the quality of the grown carbon nanotubes [9]. In this case we have a large D/G value indicating low quality tubes/fibers with a lot of defect and amorphous carbon deposition. The D/G value obtained can also be used to show the reproducibility of the process. In Figure 6 (right), this D/G value is shown for all three sets of experiments and the reference experiment with no control. It clearly shows the good reproducibility of the feedback control.

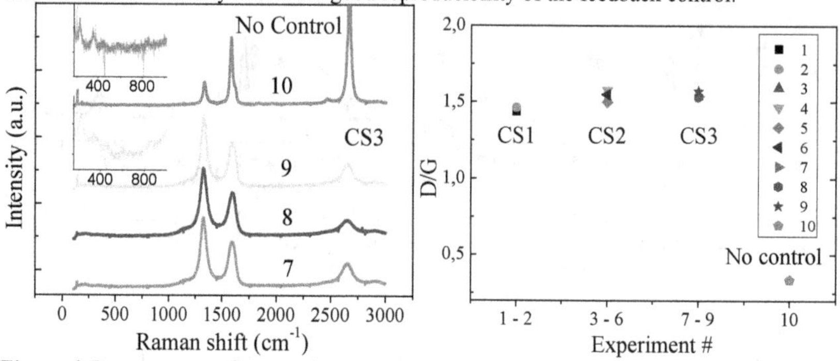

Figure 6. Raman spectra for experiments 7 – 9 of CS3 and #10 with no control (left). D/G ratio's for all control strategies as well as the uncontrolled reference.

CONCLUSIONS

In this paper a feedback controlled laser assisted chemical vapor deposition technique for the growth of carbon nanotubes was developed and demonstrated. From SEM analysis and Raman spectroscopy it was shown that a good reproducibility was achieved and that controllability is greatly enhanced.

REFERENCES

1. M. Berggren, D. Nilsson, and N. D. Robinson, *Nature Materials* **6**, 3-5 (2007).
2. K. J. Allen, *Proceedings of the IEEE* **93**, 1394-1399 (2005)
3. J. van den Brand, R. Kusters, M. Barink, and A. Dietzel, *Microelectronic Engineering* **87**, 1861-1867 (2010)
4. Y. Xu and A. Srivastava, *Int. J. Circ. Theor. Appl.* **38**, 559–575 (2010).
5. M. Haluska, Y. Bellouard, Y. van de Burgt, and A. Dietzel, *Nanotechnology* **21**, 7 (2010).
6. A. V. Melechko, V. I. Merkulov, T. E. McKnight, M. A. Guillorn, K. L. Klein, D. H. Lowndes, and M. L. Simpson, *Journal of Applied Physics* **97**, 041301 (2005)
7. R. Alexandrescu, A. Crunteanu, R.-E. Morjan, I. Morjan, F. Rohmund, L. K. L. Falk, G. Ledoux, and F. Huisken, *Infrared Physics & Technology* **44**, 43-50 (2003)
8. J. B. Park, S. H. Jeong, M. S. Jeong, S. C. Lim, I. H. Lee, and Y. H. Lee, *Nanotechnology* **20**, 185604 (2009)
9. M. S. Dresselhaus, G. Dresselhaus, R. Saito, and A. Jorio, *Physics Reports* **409**, 47 - 99 (2005)
10. R. Saito, A. Jorio, A. G. S. Filho, G. Dresselhaus, M. S. Dresselhaus, A. Grüneis, L. G. Cançado, and M. A. Pimenta, *Jpn. J. Appl. Phys.* **41**, 4878-4882 (2002)

Mater. Res. Soc. Symp. Proc. Vol. 1365 © 2011 Materials Research Society
DOI: 10.1557/opl.2011.1224

Laser Welding of Transparent High Performance Polymer Foils by using Silver Nanoparticles as Absorption Layer

Matthias Neuber and Andreas Heilmann
Fraunhofer Institute for Mechanics of Materials IWM, Walter-Hülse-Straße 1, 06120 Halle (Saale), Germany

ABSTRACT

Laser welding of transparent high performance polymer foils requires an additional absorption layer at the interface of both foils. This paper demonstrates that metallic nano-particles, e.g. gold, silver or copper, can act as such an absorption layer. Silver nanoparticles were deposited on the surface of 200 μm thick ethylene tetrafluoroethylene (ETFE) polymer foils by evaporation processes or by magnetron sputtering. For their additional mechanical stabilization, thin films produced by plasma polymerisation of hexamethyldisilazane or PTFE-polymer sputtering were deposited on top of the metal nanoparticles. Laser irradiation of the coated foil together with the untreated joining partner was performed by a continuous wave diode laser at a wavelength of 808 nm. With the defocused laser, the foils were welded and finally a nearly transparent welding seam was achieved. The nanostructure and the optical properties of the nanoparticle layer before laser irradiation were determined and compared with the nanostructure and the optical properties of the polymer metal nanocomposite after laser welding.

INTRODUCTION

Laser welding of polymers is a well-established process in polymer assembly and construction. The laser welding of two transparent polymers requires an additional absorption layer between the joining polymers parts. For these layers, special dye molecules which were developed only for these applications can be used [1]. The application of the dye molecule dispersions with polymeric components can be realized by various wet coating technologies, e.g. spraying or brushing with controlled drying. But, laser welding requires strong reproducibility and homogenous absorption of the coating which is hard to achieve. Further, the incorporation of the dye molecules and the surrounding polymeric coating into the welding seam can result in reduced long-time stability of the welding seam. Therefore, metallic nanoparticles were used as absorber for laser welding of polymer foils.
Noble metal nanoparticles with a suitable particle size and shape can be easily deposited by different vacuum deposition techniques seem to be a promising alternative absorber layer.

The optical properties of noble metallic nanoparticles embedded in polymer matrices are determined by their size and shape distribution, the spatial arrangement as well as on the surrounding media [2, 3]. The optical properties of these nanocomposite materials are a mixture of the properties of the basis components. If the metallic particles are embedded in an optical transparent media the optical properties are mainly dominated by these nanoparticles [4], [5], [6].

The noble metals gold, silver and copper are metals demonstrating plasma resonances absorption in the visible spectral range. In the following, we describe a laser welding technique, using silver nanoparticles as absorber layer to join two 200 μm thick high performance polymer foils (ethylene tetrafluoroethylene – ETFE foils).

EXPERIMENTAL

The silver nanoparticles were deposited by magnetron sputtering at a pressure below 10^{-3} Pa (high vacuum) on polymer foils of ethylene-tetrafluoroethylene with a thickness of 200 μm, a length of 15 cm and a width of 2.5 cm. In fact of the process parameters, the silver atoms grow on the samples (Volmer-Weber-growing) and form nanoparticles. Due to different positions in the reactor chamber, various nanoparticles layer can be achieved. Thus, layers with a gradual array of the metal content (polymer sample out of the center of the magnetron) and with homogeneous metal area filling factor (polymer foil direct in the center of the deposition direction of the magnetron) were produced. The Samples with homogeneous metal content allow investigations of different laser parameters and samples with gradient of the metallic nanostructure allow studies of the optical properties, e.g. the absorption behavior of the incident laser, as a function of the particle arrangement. Additional to the polymer-metal-composite layer, a layer with polymer-metal-plasmapolymer structure were created. Thus, a polymer layer with a thickness of about 50 nm was deposited by plasma polymerization of the monomer hexamethyldisilazane (HMDSN) on the top of the nanoparticles. This additional and optical transparent protection layer stabilizes the particles and increases mechanical stability.

The structure and arrangement of the silver nanoparticles were investigated by transmission electron microscopy (TEM) before and by scanning electron microscopy (SEM) and optical microscopy after the laser radiation process. To analyse the microstructure before the radiation, a special sample preparation is necessary. Therefore, a plasma polymer layer and the according particles were deposited on a pre-deposited potassium bromide or sodium chlorine layer made by vacuum evaporation. Afterwards, the polymer-metal layer was detached in distilled water and finally collected with a standard copper grid [3]. TEM investigations were performed with the Philips CM 20 with an acceleration voltage of 15 keV. Determining the area filling factor of the metal and the optical image analysis were performed with the software Analysis Pro 3.1.

The optical properties of the polymer metal composite layer before and after the irradiation were determined by a double beam spectrometer (Specord 210, Analytic Jena) between 350 and 1100 nm.

Joining two foils together, a coated polymer foil and an uncoated ETFE foil were brought into contact and irradiated in perpendicular incidence. The laser welding processes were performed with a continuous wave diode laser at a wavelength of 808 nm, an energy of 50 W, a spot diameter of 4 mm and a velocity of 0.7 m/min.

Dark field microscopy was performed by an Olympus BX 51 to investigate the polymer after the laser radiation. To analyse the mechanical properties of the welding seam, mechanical tensile strength tests were performed on strip samples by using a universal testing machine (Zwick).

RESULTS AND DISCUSSION

The figure 1 in the middle shows the silver nanoparticle layer made by sputtering on a high performance ETFE foil with increasing area filling factor from left to right.

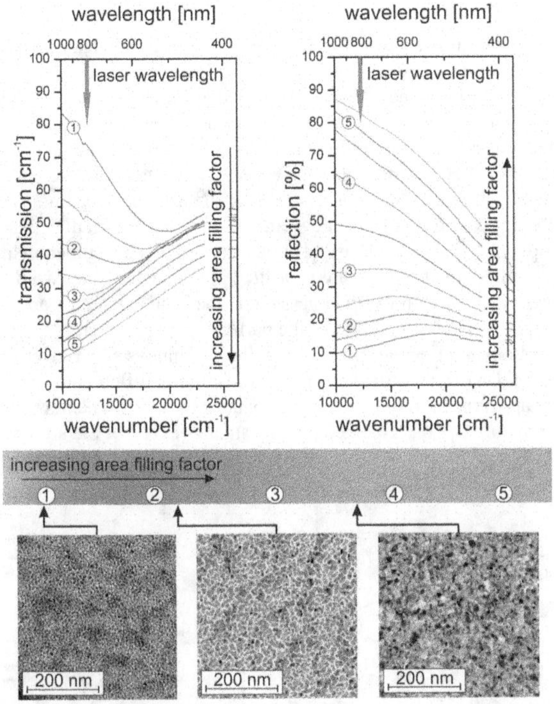

Figure 1: Optical transmission and reflection spectra of the polymer-metal-composite layer at different metal area filling factors with the according microstructure

The optical gradient results from the different nanoparticle size and shape. At different positions on this gradient, transmission and reflection spectra were performed. These spectra show in the laser wavelength range of 808 nm distinctive resonance behaviors and decreasing transmission as well as decreasing reflection behavior with increasing area filling factor. At three different positions of this composite layer, TEM micrographs depict the morphology of the particle layer. Taken in account, that the particle distribution is in plane, the description of the nanostructure can be made by using the area filling factor f_A. This factor is the relation between the area, covered with particles in comparison to the total film area.

In figure 1, f_A was determined at $f_A = 0.54$, $f_A = 0.73$, and $f_A = 0.97$ for the three different micrographs. Additional to the filling factor, the micrographs demonstrate the Volmer-Weber-growth behavior. Out of the optical spectra (figure 1), the absorption behavior of the samples was calculated for the laser wavelength of 808 nm and correlated with the nanostructure (e.g. the area filling factor) depending on the position of the gradient.

The changes of the optical properties of the irradiated areas are demonstrated in the lower part of figure 2. Figure 2 shows a picture of the welded foil with the welding seam in the middle, the light microscope pattern of the unwelded and the welded film and the according optical spectra at an area filling factor $f_A = 0.73$. This figure 2 demonstrates the changes of the optical properties of the sample during the laser radiation. Before the radiation, the absorption of the composite layer results from the plasmon resonance of the nanoparticles. The optical spectra after laser welding, demonstrates the disappearance of this resonance behavior. Thus, the polymer dominates the optical properties of the welding seam and the irradiated area, what is demonstrated in the optical spectra of figure 2. The reasons for this phenomenon are various.

Before laser radiation, the particles are deposited homogeneous and two dimensional (figure 2). Thus, the relation between the thickness d and the spatial extend x and y is approximately $x \approx y >> d$ and the orientation of the particles is perpendicular to the incident light. During the irradiation and the polymer melting, the continuous silver layer disappears and small silver floes occurs. This new particle arrangement is totally different to the unirradiated layer and prevents the appearance of the plasmon resonance behavior. The new particle orientation in the polymer matrix has changed and the smaller axis d of the layer is statistical orientated to the incident field vector (figure 2 right).

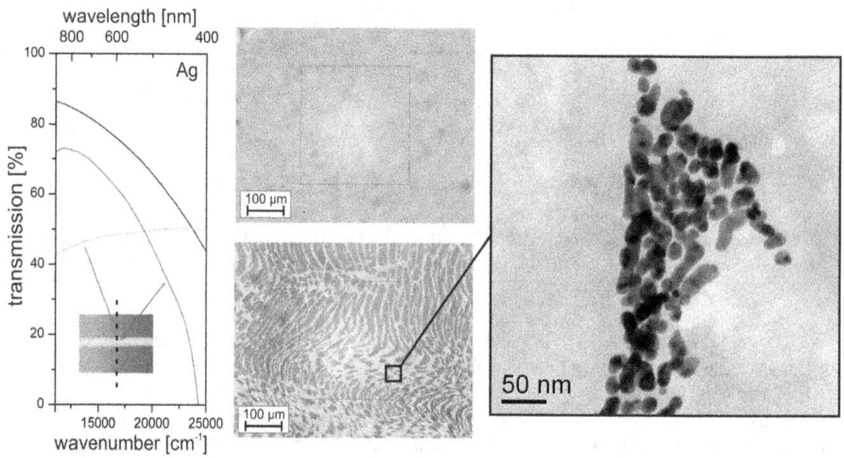

Figure 2: Optical properties and micro/ nanostructure of the welding seam

These changes in the structure of the metallic particles and the new three dimensional orientation becomes visible in the microscope (lower light microscopic picture in figure 2). Due to the thickness of the welding seam of 400 µm, the new decoration of silver islands cannot be observed directly by TEM. Thus, cross-sectional prepared samples made with a cryo ultra microtome are still necessary, getting detailed information of the new particle orientation and arrangement.

After microstructural and optical investigations before and after laser radiation welding of polymer metal composite layer, mechanical tests of the welding seams were performed. Strip samples with a geometry of 10 x 100 mm were investigated. Between the sample positions of $x = 20$ mm up to $x = 90$ mm (figure 2), the measured tensile strength was $R_m \approx 30$ N/mm². Tensile strength tests of other regions deliver no stable welding seams, thus $R_m = 0$ N/mm². This tested tensile strength of these welding seams is lower than that of conventional thermal welding seams with a maximum tensile strength of $R_m \approx 42$ N/mm² but almost higher than the upper yield point of a 200 µm thick ETFE foil ($R_m \approx 22$ N/mm²). Thus, the created polymer laser welding seams are sufficient for technical applications.

CONCLUSIONS

In the presented paper, it has been demonstrated that silver nanoparticles can act as absorption layer for laser radiation welding of optical transparent high performance polymers. Due to the plasmon resonance, the metal particles absorb laser energy during the irradiation process. The following heating of the interface between two joining partners effects the welding process of the foils. During the laser irradiation, the microstructure and the optical properties of the welding seam changes and thus, the welding seam becomes transparent. The resulting mechanical properties of these welding seams fulfill the requirements for technical applications. In addition to silver, similar results of this welding process were found for the other noble metals copper and gold.

Mechanical tensile tests of the laser welding seams have demonstrated that their tensile strength is comparable to conventional thermal welding seams. As further work, detailed TEM investigations, thermodynamic descriptions of the welding process and a description of the time dependence of the nanoparticle changes are necessary.

Due to the great flexibility of laser welding comparable to conventional thermal welding, the presented welding technique is of great interest for membrane buildings and other uses of the high-performance ETFE foil. Therefore, ongoing long time stability testing and more detailed investigations of the microstructure of the welding seam are underway.

ACKNOWLEDGMENTS

The authors thank L. Berthold for the TEM investigations, J. Lucas for support with the laser welding and T. Hanke for mechanical testing. The work was performed with financial support of the Fraunhofer Society. Finally, the authors thank the International Max Planck Research School for the support.

REFERENCES

[1] A. Boehm, M. Sieffert, T. Renner, (2007), Plastics, Additives and Compounding 7, 34-35.
[2] U. Kreibig; M. Vollmer, (1995), Optical Properties of Metal Clusters, Springer, Berlin
[3] A. Heilmann, (2003), Polymer Films with Embedded Metal Nanoparticles, Springer, Berlin
[4] A. Kiesow, S. Strohkark, K. Löschner, A. Heilmann, A. Podlipensky, A. Abdolvand, G. Seifert, (2005), App. Phys. Lett. 86, 153111.
[5] K. Loeschner, A. Kiesow, A. Heilmann, (2007), Adv. in solid state physics 46, 69-82.
[6] K. Loeschner, G. Seifert, A, Heilmann, J. Appl. Phys. 108, 073114 (2010)

Laser-Material Interactions at Micro/Nanoscales

Mater. Res. Soc. Symp. Proc. Vol. 1365 © 2011 Materials Research Society
DOI: 10.1557/opl.2011.1317

Laser-assisted surface engineering of thin film electrode materials for lithium-ion batteries

Wilhelm Pfleging [1], Robert Kohler [1], Steffen Scholz [2], Carlos Ziebert [1], Johannes Proell [1]

[1]Institute for Applied Materials, Karlsruhe Institute of Technology, Karlsruhe, Germany
[2]Manufacturing Engineering Centre, School of Engineering, Cardiff University, CF24 3AA, UK

ABSTRACT

Electrode thin films made of $LiCoO_2$, Li-Mn-O and SnO_2 were synthesized by rf magnetron sputtering on silicon and stainless steel substrates. In order to increase the active surface direct laser structuring methods using ns- and ps-laser sources were applied. A laser system operating at a wavelength of 248 nm with a pulse length of 4-6 ns and repetition rates up to 500 Hz enabled the formation of high aspect ratio micro- and sub-micron structures with feature sizes down to less than 400 nm. Subsequent to the laser structuring process, laser annealing of $LiCoO_2$ and Li-Mn-O was performed in order to achieve an appropriate crystalline phase which shows improved electrochemical cycling performance. Laser annealing was applied via a high power diode laser system operating at a wavelength of 940 nm. In case of $LiCoO_2$ the high temperature phase was obtained through laser-annealing while for Li-Mn-O the spinel phase was formed. For both $LiCoO_2$ and Li-Mn-O thin films appropriate annealing parameters were temperatures of up to 680 °C and an annealing time of 100 s.

INTRODUCTION

There is an increasing demand for powerful rechargeable battery systems suitable for automotive applications as well as for temporary storage of electrical energy provided by wind generators or solar power plants. The demand for further miniaturization has also opened the research field for micro-batteries which can be integrated into nearly any mobile device [1-3]. With respect to the type of application different challenges and requirements concerning the type and architecture of battery materials will arise. Energy density, power density and battery lifetime (cycle stability) strongly depend on the scenario of application. Nowadays, research and development of those specialized battery systems is mainly focused on powerful rechargeable lithium-ion batteries (LIB) due to their high reversible capacity [4]. Hereby, the development of cathode and anode materials plays an important role in battery technology [5]. Three dimensional battery systems have been proposed to enhance the power density while maintaining high energy densities for micro-battery systems [6]. To create thin film electrodes applicable for this purpose, different manufacturing methods have been investigated, e.g. chemical vapor deposition [7-9], pulsed laser deposition [10,11] or rf magnetron sputtering [12-15]. Investigations have been carried out to create 3D micro-batteries, yet most efforts have been focused on structuring the substrate which was silicon in most cases [16,17]. The deposition of such thin films on structured substrates is still a difficult task [18].

Current studies are focused on nano-composite electrode materials which lead to an increase in power density primarily due to large electrochemically active surfaces. In this work, further increase of active surfaces as well as a reduction of mechanical degradation was obtained by laser surface engineering of electrode materials. This is rather a new research field which was firstly applied for the cathode material lithium cobalt oxide ($LiCoO_2$) [14,19-21]. $LiCoO_2$ is one of the most commercially used LIB cathode materials. In future, this material will be replaced because of comparably high costs and the toxicity of cobalt. Hence, laser processing of advanced cathode materials such as lithium manganese oxide (Li-Mn-O) was applied. As anode material tin oxide (SnO_2) was investigated due to its enormous theoretical specific capacity of about 791 Ah/kg in comparison to graphite which is the standard anode material with a theoretical specific capacity of about 372 Ah/kg [22].

The influence of the laser treatment on the electrode materials was studied with Raman spectroscopy and scanning electron microscopy. The formation of electrochemically active and inactive phases was discussed. Finally, to investigate the electrochemical properties, the manufactured electrodes were cycled against lithium metal using standard liquid electrolyte. Laser processing parameters leading to a significant improvement of battery performance in comparison to unstructured materials were identified.

EXPERIMENTAL DETAILS

Laser technology

Different types of lasers were applied for laser processing of battery materials: Excimer laser, mode-locked Nd: YVO_4 laser and high power diode laser. The excimer laser (ATLEX-500, ATL Lasertechnik GmbH) operates at a wavelength of λ=248 nm and enables repetition rates up to 500 Hz. The laser pulse length was 4 ns (FWHM). A beam homogeneity or "flat top" profile with intensity fluctuation better than 5% was established [23]. The mode-locked laser source is a solid state (DPSS) laser from Lumera Laser GmbH (Super Rapid) and operates at a wavelength of 355 nm and delivers an average output power of 0.3 W with a pulse width of 12 ps [24]. Therefore, a high beam quality (M^2<1.2) and repetition rates up to 600 kHz can be obtained. For laser annealing a high power diode laser (FLS IronScan, Fisba Optik AG) with a maximum laser output power of 50 W and a wavelength of 940 nm was used. Via deflection mirrors the laser focus (diameter 1 mm) is scanned over the sample surface at speeds up to 2000 mm/s. The temperature in the heat-affected-zone is measured on-line during the annealing process with a pyrometer that is linked with the laser power management system. The process temperature can be controlled in the range of 120°C up to 700°C.

Electrode films

Thin films were deposited in a Leybold Z550 coating facility. Thin film electrodes were deposited by non-reactive rf magnetron sputtering from a $LiCoO_2$, SnO_2 or $LiMn_2O_4$ target in a pure Ar plasma. The process was carried out without any substrate bias voltage. A detailed description of the coating facility can be found in [25]. For the SnO_2 target the power was set up to 100 W and the total gas pressure to 1 Pa. In case of $LiMn_2O_4$ the target power was set to 200 W during deposition and the working gas pressure was 10 Pa, respectively. The $LiCoO_2$ target power was 200 W and the argon gas pressure was adjusted to 10 Pa. All coatings had a thickness of about 2-3.5 μm and were deposited on silicon wafers or stainless steel substrates. Before thin film deposition on steel, an Au buffer-layer (thickness in the range of 100 nm) was applied to the substrates for better conductivity and contact of the thin film to the substrate. The substrates were ultrasonically cleaned in acetone and isopropanol each for 15 min and were plasma-etched in argon for 15 min.

Analytics

Structural information of the coatings was obtained by micro-Raman spectroscopy at room temperature using a Renishaw-1000 system equipped with an argon ion laser (excitation wavelength $\lambda_{Raman} = 514.5$ nm, power output 21 mW).

For electrochemical characterization, the electrode films were integrated into Swagelok® cells and battery cycling was performed using an Arbin Instruments BT2000 testing system. The manufactured thin film cathodes and anodes were cycled against lithium metal using liquid electrolyte made of ethylene carbonate (EC), dimethyl carbonate (DMC) and 1M lithium hexafluorophosphate ($LiPF_6$). The charging and discharging current was chosen to match a specific "C-rate" when assuming a theoretical specific reversible capacity of 148 mAh/g for Li-Mn-O, 140 mAh/g for $LiCoO_2$ and 782 mAh/g for SnO_2.

The investigation of the surface morphology of the as-deposited, laser annealed and laser structured thin films was performed using a scanning electron microscope (Philips XL 30S).

DISCUSSION

Laser structuring of electrode materials

Direct patterning via chromium/quartz mask techniques of $LiCoO_2$, Li-Mn-O thin film cathodes as well as of SnO_2 thin film anodes was performed using ns-laser radiation with a wavelength of $\lambda = 248$ nm. Laser patterning was achieved by using a chromium/quartz mask including a grid pattern. The grating was set-up on the mask with a constant pitch distance of 40 μm in each direction. With respect to this fact, spherical structures could be formed (Fig.1) via demagnifying the mask structures by a factor of 1:10 down to the thin film surface. The aim was to form spherical surface structures with direct laser processing routes which enhance the active surfaces of the cathode materials $LiCoO_2$ and Li-Mn-O and therefore allow better capacity retention during cycling of the cathode which has already been shown by Kohler et al. [19] for $LiCoO_2$ and by Proell et al. [26] for Li-Mn-O thin films sputtered at 0.25 Pa argon pressure. In case of SnO_2, the goal was to achieve large active surfaces for better Li-ion diffusion as well as to apply an appropriate surface pattern which can compensate the extremely high volume changes of the bulk material during electrochemical cycling [27]. Actually, material removal was processed down to the substrate for all materials which is demonstrated in figure 1a-c. Therefore, laser parameters such as energy density ε, pulse number N and repetition rate ν_{rep} had to be adjusted to guarantee high aspect ratio structures with small distances in between. This is not only necessary to reduce material removal down to a minimum but also to enlarge the active surface area up to a maximum. Surface patterns with small distances of below 200 nm (figure 1c) between each spherical structure could be formed by laser ablation of SnO_2 thin films. The cone-shaped structures remain nearly debris-free in case of $LiCoO_2$ and SnO_2 while for Li-Mn-O debris could be detected after laser structuring under ambient air. Thermal impact occurred within ns-laser structuring for all structures as was proofed via SEM images.

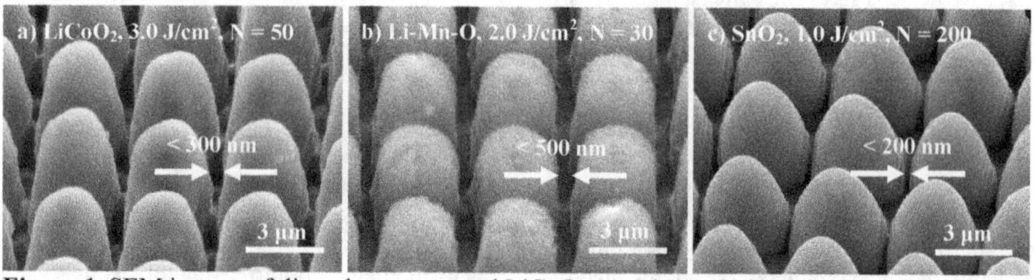

Figure 1. SEM images of direct laser patterned $LiCoO_2$ (a), Li-Mn-O (b) cathodes and SnO_2 (c) anode thin films as electrode materials for lithium-ion batteries. Laser structuring was performed using UV-laser radiation with $\lambda = 248$ nm operating with ns-pulses. All structures were formed using ambient air as processing atmosphere. Laser repetition rate during structuring process was set to (a) $v_{rep} = 200$ Hz, (b) $v_{rep} = 100$ Hz and (c) $v_{rep} = 100$ Hz, respectively.

Ultrashort pulse ps-laser structuring of SnO_2 anode thin films was investigated with $\lambda = 355$ nm in a first step in order to evaluate appropriate parameters for material removal down to substrate and in order to avoid thermal influence on the surrounding bulk material. Different line energies L were tested while a line energy of L = 0.5 J/m removes only half of the film thickness which could be determined to 3.5 µm for all SnO_2 thin films. Actually, in figure 2, a-c the contours of the structured regions were visualized. Figure 2a shows clearly the Gaussian laser profile while in figure 2b material removal could be applied down to substrate and a change in cross section profile in the lower part of the thin film was observed probably due to a change in ablation characteristics in the interface of SnO_2 thin film and silicon substrate. A line energy of L = 1.0 J/cm seemed to be an appropriate value for direct laser structuring of SnO_2 thin films down to substrate with a height of about 3.5 µm (figure 2b). Higher line energies, as in the case of figure 1c, show also partial substrate removal and melted regions between the removed material and the surrounding bulk.

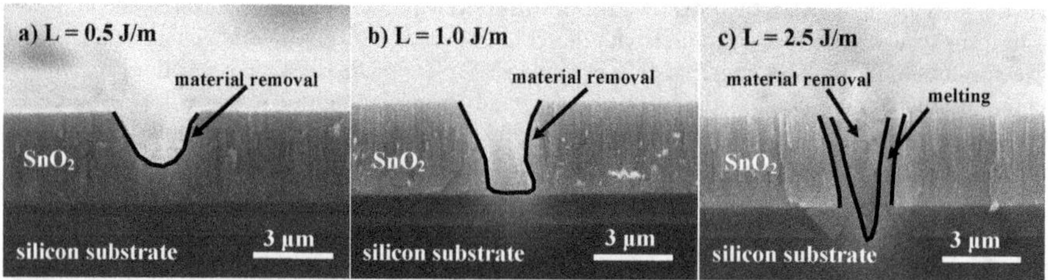

Figure 2. SEM cross-sectional images of ps-laser structuring processes of as-deposited SnO_2 thin films with defined material removal using line energies of (a) L = 0.5 J/m, (b) L = 1.0 J/m and (c) L = 2.5 J/m. Laser structuring was performed under ambient air using a constant repetition rate of $v_{rep} = 50$ kHz. The different line energies were applied using a constant laser power of P = 5 mW and varying the laser scan speed.

The laser-induced channel depth d was measured as a function of the line energy L in figure 3. The laser power was varied in the range of 5 mW up to 25 mW. The SnO_2 film thickness was 3.5 µm. Within a laser power of P = 5 mW six channel depths d were measured for different line energies L. They show a linear behavior (figure 3) and the channel depth varies between 2 and 3.5 µm in the range of L = 0.5 – 1.0 J/m. Within a power of P = 10 mW three channel depths (d = 2.72 – 3.39 µm) were measured also indicating a linear progress between L = 1.0 – 1.25 J/m with a channel depth d of ~3.39 µm at 1.25 J/m which is near to the substrate due to a film thickness of 3.5 µm. Applying a line energy of 1.5 J/m by using P = 15 mW results in a channel depth of d = 3.25 µm. All ablation experiments were carried out under ambient air using a constant repetition rate of v_{rep} = 50 kHz. It seems obvious, that the initial laser power P plays an important role during structuring due to the fact that comparison of channel depths for specific line energies does not consequently lead to the same value of channel depth d when assuming further linear distribution of the fitted data points for each power.

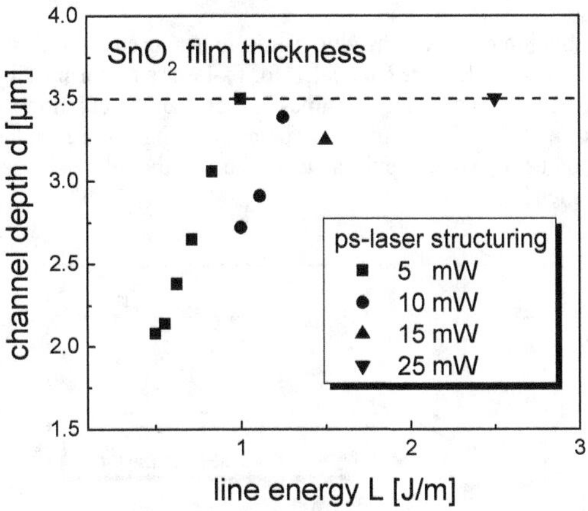

Figure 3. Control of channel depth d of ps-laser structured SnO_2 thin films in dependency of the line energy L within laser powers of P = 5 mW, P = 10 mW and P = 15 mW. Ps-laser structuring was performed using a constant repetition rate v_{rep} = 50 kHz by varying the laser scan speed. All ablation experiments were carried out under ambient air.

For a laser power of 25 mW and a scan speed of 10 mm/s complete removal of the SnO_2 layer could be observed. The silicon substrate remained undamaged and a nearly rectangular cross section profile was obtained (figure 4a). Good surface qualities were also obtained by using ns-laser radiation (figure 4b). Nevertheless, it is obvious that ns-laser radiation induced thermal impact into the SnO_2 material leading to burr formation at the sidewalls of the channel. The amount of debris was reduced significantly by using helium as processing gas (figure 4b) in comparison to laser processing in ambient air (figure 4a).

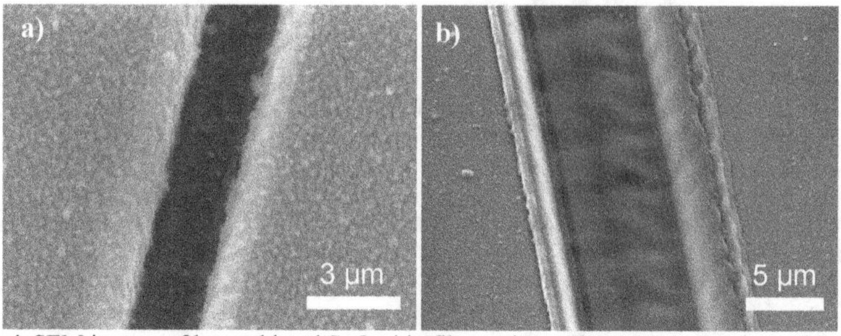

Figure 4. SEM images of laser ablated SnO_2 thin films using ps-laser (a) and ns-laser (b). (ps-laser ablation: power 25 mW, scan velocity 10 mm/s, repetition rate 50 kHz; ns-laser ablation: 1.9 J/cm^2, repetition rate 100 Hz).

In order to compare the ablation rates R of ns- and ps-laser structured SnO_2 thin films, the energy density as well as the ablation rate R were calculated for ps-laser structuring. Figure 5 shows that the ablation rate R using ps-laser is two to three times higher than the ablation rate R for ns-laser structuring in the range of $\varepsilon = 1.4 - 2.8$ J/cm^2. Laser ablation using ps-laser radiation is more efficient than laser ablation using ns-laser pulses due to the fact that ultrashort laser material processing leads to a smaller formation of heat losses.

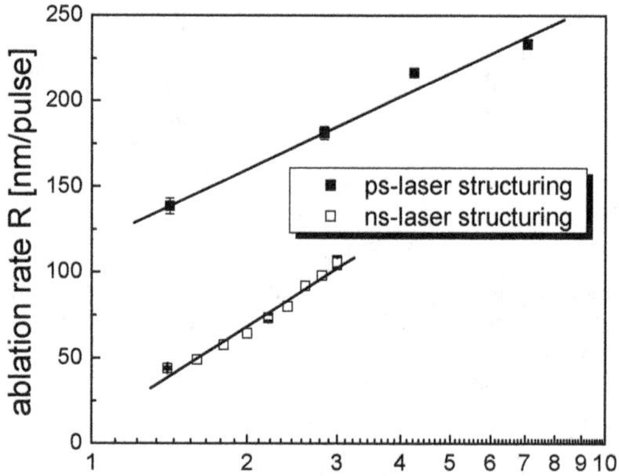

Figure 5. Ablation rate R as function of laser fluence for ps- and ns-laser radiation. The ps-laser structuring was performed using a constant repetition rate $\nu_{rep} = 50$ kHz by varying laser scan speed and ns-laser structuring was performed using $\lambda = 248$ nm and $\nu_{rep} = 100$ Hz.

Laser annealing of thin film cathodes

The cathode material $LiCoO_2$ as well as lithium manganese oxide compounds require subsequent heat treatment processes after thin film deposition on steel substrates in order to form a crystalline and electrochemical active phase. In this research study the active battery phases could be formed for $LiCoO_2$ as well as for Li-Mn-O thin films. In case of $LiCoO_2$ laser-annealing for 100 s at 680 °C obviously leads to the high temperature (HT-$LiCoO_2$) phase (figure 6b). This phase is strongly characterized by two sharp Raman peaks at 486 cm^{-1} (E_g mode) and 595 cm^{-1} (A_{1g}) indicating the bending (E_g) and stretching (A_{1g}) vibrations of the cobalt-oxygen molecules. For laser-annealing of Li-Mn-O thin films under ambient air a Raman spectrum was detected which is very similar to the spectra which has already been recorded by Julien et al. [28] for spinel $LiMn_2O_4$ (figure 6d). Spinel-like Li-Mn-O compounds have the feature of being relatively weak in their Raman activity because of their polaronic character. Three Raman peaks can be assigned to the spinel structure out of figure 6d having their maximum at 483, 583 and 625 cm^{-1}. The broad peak at 625 cm^{-1} (A_{1g} symmetry) can be assigned to a stretching vibration of manganese and oxygen compounds within the MnO_6 octahedra [29] while the low intensity shoulder at 583 cm^{-1} (F_{2g} species) is supposed to be related to the averaged manganese oxidation state, taking into account that within the spinel structure isotropic $[Mn^{4+}]O_6$ and Jahn-Teller distorted $[Mn^{3+}]O_6$ octahedra exist at the same time. The weak band at 483 cm^{-1} can also be assigned to a F_{2g} species (νMn-O) of the spinel phase [29].

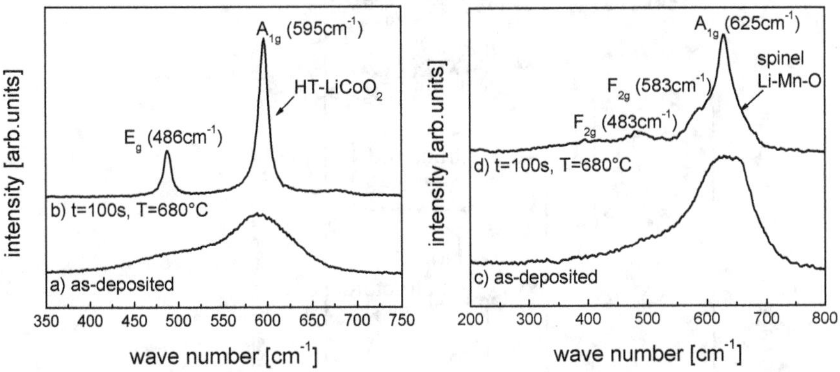

Figure 6. Raman spectra of as-deposited and laser-annealed $LiCoO_2$ (left) and Li-Mn-O (right) thin films. Laser-annealing was performed for t = 100s at T = 680 °C under ambient air. Both, the $LiCoO_2$ and the Li-Mn-O thin films were deposited on stainless steel substrates for annealing experiments (10 Pa argon pressure during deposition).

While Raman measurements do not consequently show that a large fraction of the thin film volume and therefore the bulk material has crystallized, X-ray diffraction measurements were investigated which show that also bulk crystallization occurred due to the appearance of characteristic reflexes for both, hexagonal HT-$LiCoO_2$ and cubic spinel-like Li-Mn-O. In case of Li-Mn-O, laser-annealed thin films were tested via cyclovoltammetry using very slow sweep rates of 0.02 mV/s. They revealed sharp redox-peaks around the 4V-region and therefore are in good agreement with those being estimated for spinel $LiMn_2O_4$ which also indicate a well-crystallized thin film. Finally, figure 6a,c show broad Raman spectra for as-deposited thin films which can not be easily assigned to one specific phase because of inhomogeneity in local structure or amorphous states after the deposition process. It is obvious that laser-annealing is required for adjacent battery testing. This was already demonstrated in [19,26].

Electrochemical cycling

Electrochemical testing was performed for the unstructured and structured electrode thin film materials made of SnO_2, $LiCoO_2$ and Li-Mn-O using the standard liquid electrolyte composed out of a solvent EC/DMC 1:1 with 1M LiPF6 as electrolyte salt. A lithium metal foil was used as the anode and a Swagelok® test cell was used. When describing batteries, the charge/discharge current is often expressed as a C-rate in order to normalize against battery capacity. A C-rate is a measure of the rate at which a battery is discharged relative to its maximum capacity. A "1C" rate means that the discharge current will discharge the entire battery in 1 hour.

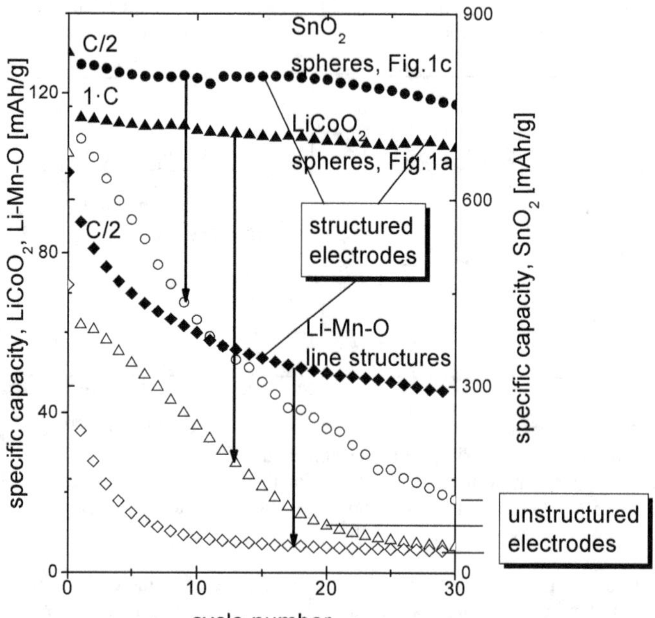

Figure 7. Specific discharge capacities of ns-laser structured and unstructured electrode materials. Electrochemical cycling was performed against Li metal foil applying a constant dis-/charging current of 1·C for $LiCoO_2$ and C/2 for SnO_2 and Li-Mn-O (Swagelok® test cell).

The filled symbols indicate the structured electrode materials and the unfilled symbols represent the corresponding unstructured electrode thin films. More details are given in the text.

For SnO_2 cycling was performed in between the voltage window of 0.02 V to 1.2 V at a dis-/charging rate of C/2. It was calculated, that with respect to this conditions, a maximum theoretical specific capacity of 791 mAh/g is possible. From figure 7 it can be seen, that this theoretical value can be maintained for nearly 30 cycles but only for the laser structured SnO_2 thin film while the unstructured thin film shows a smaller capacity and a strong decrease of capacity as function of cycle number. The corresponding surface structure of the laser processed thin film is shown in figure 1c. The loss in capacity when cycling the unstructured SnO_2 thin film results from the huge change in volume during cycling attended by the loss of electrical contact of thin film and substrate [22]. Laser structured surfaces enable a better compensation of volume changes in the films during electrochemical cycling while free-standing structures are most preferable. For instance micro holes produced in SnO_2 thin films showed no improvement of capacity retention and a cell performance similar to the unstructured films was obtained. Free standing structures were also produced in $LiCoO_2$ and Li-Mn-O thin films for subsequent electrochemical tests (figure 7). In case of $LiCoO_2$ structures similar to those in figure 1a were produced by direct laser ablation using chromium/quartz masks. For C-rates smaller or equal C/5 the theoretical specific capacity of 140 mAh/g can be obtained for the applied voltage window of 3 V to 4.2 V. For higher C-rates such as 1·C, the capacity decreases down to 110 mAh/g (figure 7). Nevertheless, the capacity retention is extremely stable in comparison to that of the unstructured $LiCoO_2$ thin film. After 30 cycles the cell test of the unstructured $LiCoO_2$ thin film was failed. For Li-Mn-O line structures with a pitch distance of 3.2 μm and a groove width of 400 nm were produced for battery cell tests shown in figure 7 (voltage window for Li-Mn-O: 3.0 V - 4.3 V). With respect to a compensation of volume changes during cycling this design is not the best choice. Nevertheless, even with this type of surface pattern a significant improvement of battery cell performance with respect to capacity retention at high discharge rates (C/2) is obtained in comparison to the unstructured thin film. The degradation mechanisms of Li-Mn-O electrode thin films is much more complex than for SnO_2 or $LiCoO_2$ thin films: During cycling of Li-Mn-O thin films the Jahn-Teller distortion takes place due to reduction of manganese ions which can be accompanied with a structural change and a loss of active material due to the change in oxidation state and dissolution of MnO into the electrolyte [30].

CONCLUSIONS

In order to produce free-standing microstructures such as spherical structures or ridges on $LiCoO_2$, Li-Mn-O and SnO_2 thin films UV-laser structuring was applied using ps- and ns-laser pulses. Therefore, specific laser parameters had to be applied for each material to achieve material removal down to the substrate. It has been proofed for each material that the formation of such surface structures which are known to show good electrochemical cycling behavior is possible in general. The ns-laser pulses operated at a wavelength of 248 nm and periodic structures were obtained by using chromium/quartz mask techniques. The ps-laser ablation was used in serial direct writing mode. In case of SnO_2 thin films it has been shown, that ps-laser processing is much more efficiently than ns-laser processing resulting in an ablation rate being two to three times higher. The ps-laser structuring of SnO_2 thin films were investigated, whereas a line energy of L = 1.0 J/m resulting from a laser power of P = 5 mW leads to defined material removal down to substrate. The channel depth d was measured as a function of the line energy L. It was observed that at a specific line energy L, which can be obtained for the laser powers of P = 5-15 mW does not consequently lead to the same channel depth d.

Further on, rapid laser-annealing methods were investigated for $LiCoO_2$ and Li-Mn-O thin films for setting-up a crystalline and appropriate battery phase like HT-$LiCoO_2$ or spinel-like Li-Mn-O within a short processing time span. This was achieved using high annealing temperatures of up to 680 °C to overcome the formation of inactive phases like Li_2MnO_3 as in the case of Li-Mn-O.

Laser structuring and annealing are appropriate tools for fabrication of 3D micro batteries with improved capacity retention and optimized specific capacity especially when operating at high discharge currents.

ACKNOWLEDGMENTS

We are grateful to our colleagues H. Besser and T. Dobrev for their technical assistance during laser material processing. We would also like to thank N. Thiel and J. Ye for the sputtering of the thin films. We gratefully acknowledge the financial support by the Federal Ministry for Education and Research (BMBF) in the BMBF-project 03SF0344A "Li-ion battery cells based on novel nanocomposite materials" (LIB-NANO) in the framework of "Lithium-Ion Battery LIB-2015". The ps-laser machining was carried out with the support of the European Research Infrastructure EUMINAfab program (funded under the FP7 specific program Capacities, Grant Agreement Number 226460).

REFERENCES

1. W.C. West, J.F. Whitacre, V. White, and B.V. Ratnakumar, Journal of Micromechanics and Microengineering, 12, 58 (2002).
2. L. Baggetto, R.A.H. Niessen, F. Roozeboom, and P.H.L. Notten, Advanced Functional Materials, 18, 1057 (2008).
3. J. Song, X. Yang, S.S. Zeng, M.Z. Cai, L.T. Zhang, Q.F. Dong, M.S. Zheng, S.T. Wu, and Q.H. Wu, Journal of Micromechanics and Microengineering, 19, (2009).
4. B. Scrosati and J. Garche, Journal of Power Sources, 195, 2419 (2010).
5. C. Liu, F. Li, L.P. Ma, and H.M. Cheng, Advanced Materials, 22, E28 (2010).
6. R.W. Hart, H.S. White, B. Dunn, and D.R. Rolison, Electrochemistry Communications, 5, 120 (2003).
7. H. Kobayashi, Y. Uebou, T. Ishida, S. Tamura, S. Mochizuki, T. Mihara, M. Tabuchi, H. Kageyama, and Y. Yamamoto, Journal of Power Sources, 97-8, 229 (2001).
8. P.Y. Liu, J.F. Chen, and W.D. Sun, Vacuum, 76, 7 (2004).
9. M. Kwoka, L. Ottaviano, N. Waczynska, S. Santucci, and J. Szuber, Applied Surface Science, 256, 5771 (2010).
10. N. Kuwata, R. Kumar, K. Toribami, T. Suzuki, T. Hattori, and J. Kawamura, Solid State Ionics, 177, 2827 (2006).
11. H. Xia, S.B. Tang, and L. Lu, Journal of the Korean Physical Society, 51, 1055 (2007).

12. K.F. Chiu, C.C. Chen, K.M. Lin, H.C. Lin, C.C. Lo, W.H. Ho, and C.S. Jiang, Vacuum, 84, 1296 (2010).
13. J. Song, M.Z. Cai, Q.F. Dong, M.S. Zheng, Q.H. Wu, and S.T. Wu, Electrochimica Acta, 54, 2748 (2009).
14. B. Ketterer, H. Vasilchina, K. Seemann, S. Ulrich, H. Besser, W. Pfleging, T. Kaiser, and C. Adelhelm, International Journal of Materials Research, 99, 1171 (2008).
15. C. Ziebert, B. Ketterer, M. Rinke, C. Adelhelm, S. Ulrich, K.H. Zum Gahr, S. Indris, and T. Schimmel, Surface & Coatings Technology, 205, 1589 (2010).
16. D. Golodnitsky, V. Yufit, M. Nathan, I. Shechtman, T. Ripenbein, E. Strauss, S. Menkin, and E. Peled, Journal of Power Sources, 153, 281 (2006).
17. L. Baggetto, J.F.M. Oudenhoven, T. van Dongen, J.H. Klootwijk, M. Mulder, R.A.H. Niessen, M.H.J.M. de Croon, and P.H.L. Notten, Journal of Power Sources, 189, 402 (2009).
18. A.V. Jeyaseelan and J.F. Rohan, Applied Surface Science, 256, S61 (2009).
19. R. Kohler, P. Smyrek, S. Ulrich, M. Bruns, V. Trouillet, and W. Pfleging, Journal of Optoelectronics and Advanced Materials, 12, 547 (2010).
20. R. Kohler, J. Proell, S. Ulrich, V. Trouillet, S. Indris, M. Przybylski, and W. Pfleging, Laser-Based Micro- and Nanopackaging and Assembly III, 7202, 720207 (2009).
21. R. Kohler, M. Bruns, P. Smyrek, S. Ulrich, M. Przybylski, and W. Pfleging, Laser-Based Micro- and Nanopackaging and Assembly IV, 7585, 75850O (2010).
22. W.J. Zhang, Journal of Power Sources, 196, 13 (2011).
23. W. Pfleging, M. Przybylski, and H.J. Bruckner, Laser-based Micropackaging, 6107, G1070 (2006).
24. D. Karnakis, G. Rutterford, M. Knowles, T. Dobrev, P. Petkov, and S. Dimov, Photon Processing in Microelectronic and Photonics V, 6106, 60604 (2006).
25. A. Kratzsch, S. Ulrich, H. Leiste, M. Stuber, and H. Holleck, Surface & Coatings Technology, 119, 949 (1999).
26. J. Proell, R. Kohler, C. Adelhelm, M. Bruns, M. Torge, S. Heißler, M. Przybylski, C. Ziebert, and W. Pfleging, Laser-based Micro- and Nanopackaging and Assembly V, 7921, in press (2011).
27. R. Kohler, H. Besser, M. Hagen, J. Ye, C. Ziebert, S. Ulrich, J. Proell, and W. Pfleging, Microsystem Technologies, 17, 225 (2011).
28. C.M. Julien and M. Massot, Materials Science and Engineering B-Solid State Materials for Advanced Technology, 100, 69 (2003).
29. C.M. Julien and M. Massot, Materials Science and Engineering B-Solid State Materials for Advanced Technology, 97, 217 (2003).
30. K. Edstrom, T. Gustafsson, and J.O. Thomas, Electrochimica Acta, 50, 397 (2004).

Mater. Res. Soc. Symp. Proc. Vol. 1365 © 2011 Materials Research Society
DOI: 10.1557/opl.2011.888

What Can Lasers Do in the Nano-Fabrication of Carbon Nanotube Based Devices?

Yun Shen Zhou, Wei Xiong, Masoud Mahjouri-Samani, Yang Gao, Matt Mitchell, and Yong Feng Lu[1]
Department of Electrical Engineering, University of Nebraska-Lincoln,
Lincoln, NE, 68588-0511, U.S.A.

ABSTRACT

Numerous applications based on CNTs have been conceived and developed at laboratory scale. However, only a handful of applications have been successfully implemented due to the difficulties in controlled growth, manipulation, and integration of CNTs. In spite of countless efforts having been devoted into this field, high-performance-on-demand solution packages are still absent. In this study, we investigated applications of lasers in the controlled growth and integration of CNTs, and developed laser-based strategies to achieve nano-fabrication of CNT-based devices. By making use of unique features of lasers, we achieved 1) parallel integration of CNTs into pre-designed micro/nano-architectures in a single-step laser-assisted chemical vapor deposition (LCVD) process, 2) selective removal of metallic CNTs in open air, 3) growing CNTs of controlled-alignments, and 4) diameter modulation in individual CNTs. The laser-based strategies developed in this study suggest a laser-based solution-package to meet the challenges for the nano-fabrication of CNT-based devices and promises a reliable and scalable approach to achieve CNT-integrated devices.

INTRODUCTION

Carbon nanotubes (CNTs) belong to a family of one-dimensional (1D) cylindrical carbon molecules demonstrating a rich diversity of chiral structures and remarkable properties. Enchanted by their intrinsic 1D structure and extraordinary properties, tremendous efforts have been devoted to the investigations of CNTs [1-3]. A wide spectrum of CNT-based applications has been conceived and investigated [1-3]. CNTs have been envisioned as one of the most promising candidates for fabricating nano-electronics, optics, opto-electronics, etc [4,5]. To achieve full potentials of CNTs, successful implementation of CNTs has to be accomplished. Extensive investigations have been made in fabricating CNT-based components and devices, such as transistors, logic gates, interconnects, sensors and detectors [1-5]. However, successful implementation of CNT-based devices is yet far from reality due to several critical challenges, including precise integration, stable wiring, alignment direction, and electrical type. Ideally, it is desired that CNTs of determined electrical types were precisely wired between pre-designed electrodes with stable contacts for device fabrication [6]. However, such precise control is extremely difficult due to the ultra-tiny size of CNTs, high-resolution nanoscale positioning, and subtle structure diversity. Extensive investigations have been carried to address the challenges. To achieve controlled manipulation and integration of CNTs, scanning-probe-microscope-assisted patterning, template-directed assembly, and dielectrophoresis deposition have been developed [7-10]. Monodispersed CNTs of uniform electrical types have been obtained through preferential growth of semiconducting CNTs (s-CNTs) and post-growth separation techniques [11,12]. CNTs of different alignments were acquired through E-field, gas-flow, fast-heating, and

[1] Correspondence should be addressed to Dr. Y. F. Lu, Tel: 402-472-8323, Fax: 402-472-4732, Email: ylu2@unl.edu.

template-assisted approaches [13-16]. However, existing drawbacks (such as slow processing speed, coarse control, solution contamination, and polymer/DNA wrapping) of individual methods and diversified approaches make it difficult to develop a systematic solution package for fabricating CNT-based devices.

Lasers of a wide spectrum of wavelengths have been successfully deployed in thousands of highly varied applications in every section of modern society. Compared with conventional energy sources, lasers exhibit several unparallel advantages, including spatial coherence, intense energy delivery, temporal coherence, monochromaticity, localized heating, optical near-field effect, beam polarization, and a wide spectrum of photons. The distinctive features endow laser unique strength in the fabrication of CNT-based devices. In this study, a series of laser-based approaches were developed to provide a laser-based solution-package to meet the challenges for the nano-fabrication of CNT-based devices, including 1) parallel integration of CNTs into pre-designed micro/nano-architectures, 2) growing CNT arrays of controlled alignments, 3) selective removal of metallic CNTs, and 4) growing diameter modulated CNTs. The laser-based approaches provide a new insight into the nano-fabrication of CNT-based devices, and promise a reliable and scalable approach to achieve CNT-integrated devices.

EXPERIMENT

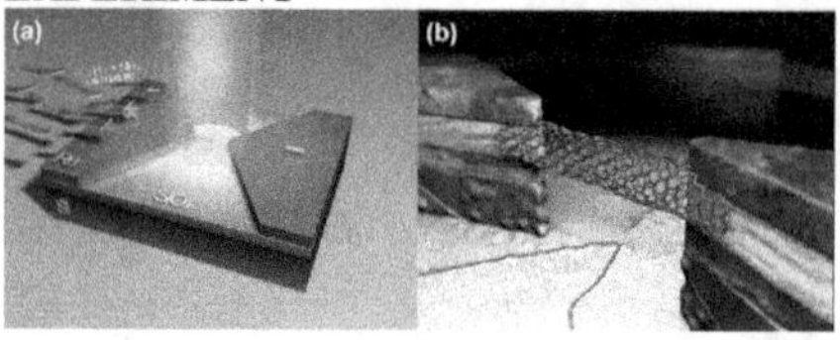

Figure 1 (a) A schematic diagram of the LCVD process and (b) a schematic diagram of a CNT-integrated bridge structure.

An LCVD system was used in this study to achieve the *in situ* growth and integration of CNTs into pre-designed nano-architectures, as shown in Fig. 1(a). SiO_2/Si substrates were used. Patterned metallic electrodes were fabricated by photolithography, sputtering, and lift-off processes. The metallic electrodes consisted of multiple metallic layers, including a ruthenium (Ru) layer (200 nm), aluminum (Al) layer (5 nm), iron (Fe) layer (1 nm), and Al layer (20 nm) sequentially from the bottom to the top. The Ru film was used as the electrode. A bimetallic catalyst containing Al and Fe was used for growing CNTs [17]. Ultra-sharp electrode tips were fabricated by focused ion beam (FIB) nano-machining using an FEI Strata 200xp FIB system. CNT growth was performed through the LCVD process, as shown in Fig. 1(a). A continuous-wave (CW) CO_2 laser (Synrad, Firestar v40) was used. The laser beam has a horizontally linearized polarization. The laser power was adjusted continuously to maintain a stable substrate temperature. An annealing treatment was carried out to activate the bimetallic catalyst [17]. The annealing process was performed at 500 °C in an ammonia atmosphere (10 Torr) for 2 min. After annealing, the substrate was protected in vacuum and cooled down to room temperature. A DC voltage (about 1 V/μm) was applied between the electrodes to direct the growth direction of CNTs. A 1 MΩ resistor was connected in the circuit to protect CNTs from being damaged by current overflow. The Si substrate was heated up to 500 °C before feeding the gaseous precursors. A gas mixture of acetylene (C_2H_2) and anhydrous ammonia (NH_3) with a volume ratio of 1:10 was fed into the LCVD chamber. A non-contact infrared pyrometer (Omega, OS3750) was used to monitor the substrate temperature. The reaction pressure was maintained at 10 Torr. The substrate temperature was controlled around 500 °C, which is almost 200 °C lower than a conventional LCVD process. A voltmeter was used to monitor the growth process of CNTs by detecting the voltage drop across the resistor. The LCVD process was terminated immediately when the circuit was closed by a CNT connection. Figure 1(b) shows a schematic structure of a CNT-bridge structure. For growing CNT arrays of different alignments, conventional metallic electrodes, including point-to-point and cross-shaped patterns, were fabricated. External electrical biases of different polarities were applied to the metallic electrodes for alignment control. For growing diameter modulated CNTs, incident laser power was adjusted instantly to achieve immediate temperature change during the LCVD growth process. CNTs were grown with a decreasing temperature profile from 650 and 450 °C in ten equal steps of 20 °C/step.

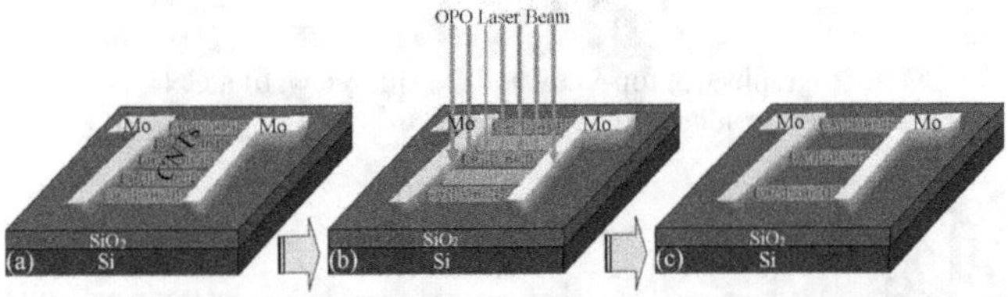

Figure 2 Schematic diagram of the experimental process for selective removing m-CNTs: (a) a CNT array before laser irradiation, (b) the laser irradiation process, and (c) a s-CNT array after laser irradiation.

Selective removal of metallic CNTs (m-CNTs) was achieved through a post-growth laser processing, as shown in Fig. 2. Both well aligned CNTs between parallel electrodes and randomly distributed CNT mixtures were prepared and used for this study. An optical parametric oscillator (OPO) wavelength tunable laser (Continuum, Panther ® EX) was used as the laser irradiation source, which provides laser emission from 215 to 2550 nm. The OPO laser was operated at a wavelength of 2 μm. Laser beams of shorter wavelengths could cause damages to both s- and m-CNTs due to the high-energy photons. Due to the sharp drop of the output power for wavelength over 2 μm, wavelengths larger than 2 μm were not investigated. The OPO laser provides laser pulses with pulse duration of 6 ns at a frequency of 10 Hz. The energy of each pulse is about 3 mJ. The power density was 1.8 MW/mm^2.

All samples were characterized by using a scanning electron microscope (Hitachi S4700 field-emission SEM), transmission electron microscope (FEI Tecnai G2 F30 field-emission TEM), semiconductor parameter analyzer (Agilent HP4155C), and micro-Raman spectrometer (Renishaw inVia dispersive micro-Raman spectrometer).

RESULTS and DISCUSSION

1. Parallel integration of CNTs into pre-designed micro/nano-architectures

To investigate parallel integration of CNTs into pre-designed micro/nano-architectrues, electrode patterns containing multiple tip-shaped electrode arrays were fabricated. Tip-shaped electrodes were used as optical antennas to focus and enhance optical fields within localized regions. A simple electrode structure containing a pair of electrode tips was used, as shown in

Fig. 3(a). The laser polarization was set to be parallel to the electrode tip arrangement. Figures 3(a) and (b) show a top-view and side-view, respectively, of a typical CNT-bridge structure. A CNT is clearly observed to bridge the electrode tips and suspend above the substrate surface. Typical Raman spectrum of the CNT-bridge, Fig. 3(c), shows a sharp radial breathing mode (RBM) shift at 207.6 cm^{-1}, a Lorentzian line-shaped G^{-} band at 1567.8 cm^{-1}, and a strong G band at 1591.5 cm^{-1} indicating the existence of s-CNTs with an estimated diameter of 1.2 nm [18-20].

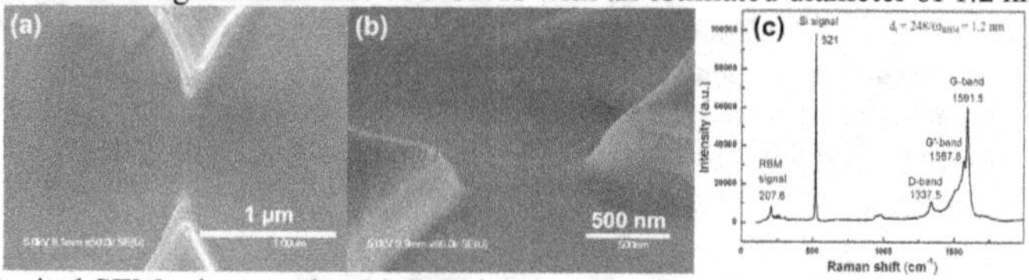

Figure 3 Typical SEM micrographs, (a) top-view and (b) side-view, of a CNT-bridge structure. (c) Raman spectrum of the CNT-bridge structure.

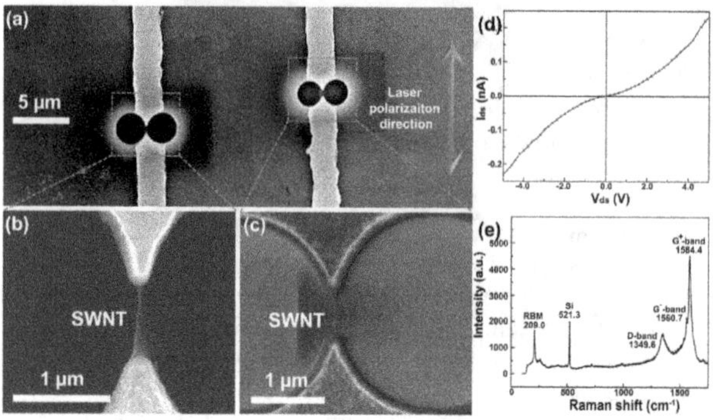

Figure 4 (a) An SEM micrograph showing the electrode pattern; (b) and (c) SEM micrographs of the squared regions in (a), showing the CNT- bridge structures; (d) *I–V* measurement of the device and (e) Raman spectrum of the CNTs.

Similar to Fig. 3(a), an electrode structure containing two pairs of parallel electrode tips was fabricated, as shown in Fig. 4(a). The laser polarization was parallel to the electrode tip arrangement, as shown in Fig. 4(a). Figures 4(b) and (c) exhibit zoomed images of the squared regions in Fig. 4(a) and show precisely integrated CNTs between each pair of the electrodes. Only one CNT was observed between each pair of the electrodes. Growth of CNTs was observed selectively at the electrode tips. A nonlinear I_{ds}–V_{ds} curve was observed at zero gate voltage, as shown in Fig. 4(d). Due to the existence of Schottky barriers formed at the CNT–electrode contacts, a low I_{ds} value was observed. In the Raman spectrum, as shown in Fig. 4(e), a sharp RBM Raman shift at 209.9 cm^{-1} and a Lorentzian line-shaped G^{-} band are observed, and indicate the growth of s-CNTs with an estimated diameter of 1.19 nm [18-20]. A more sophisticated electrode pattern containing four pairs of electrode tips were fabricated, as shown in Fig. 5(a). Two pairs of tip-shaped electrodes were fabricated. The laser polarization was set to be parallel to the arrangement of the tip arrays, as shown in Fig. 5(a). Figures 5(b) and (c) show zoomed images of the squared regions in Fig. 5(a), and clearly show precisely integrated CNTs between each pair of the electrodes. Growth of CNTs was observed selectively at the electrode tips. Most of the CNTs were precisely wired from tip to tip. Some CNTs were observed going astray, as shown in Fig. 5(c), and were ascribed to the distorted electric field distributions due to the formation of the isolated metallic island. A nonlinear I_{ds}–V_{ds} curve was observed at zero gate voltage, as shown in Fig. 5(d). According to the Raman spectrum, multiple sharp RBM peaks were identified at 194.1 and 301.3 cm^{-1}, indicating CNTs with diameters of 1.28 and 0.82 nm, respectively [18-20]. The G^{-} band at 1552.7 cm^{-1} shows a Lorentzian lineshape, indicating s-CNTs [21-23]. The semiconducting characteristics of the integrated CNTs were further demonstrated by electrical transportation characterization, including I_{ds}–V_{ds} and I_{ds}–V_{g} curves, as shown in Figs. 6(a) and (b). The CNT channels showed typical *p*-type characteristics with obvious gate voltage modulation demonstrating the semiconducting characteristics of the CNTs.

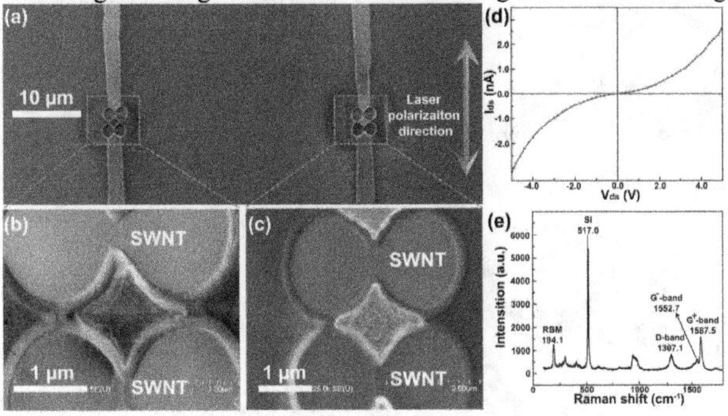

Figure 5 (a) An SEM micrograph showing the electrode pattern; (b) and (c) SEM micrographs of the squared regions in (a), showing the CNT- bridge structures; (d) *I*–*V* measurement of the device and (e) Raman spectrum of the CNTs.

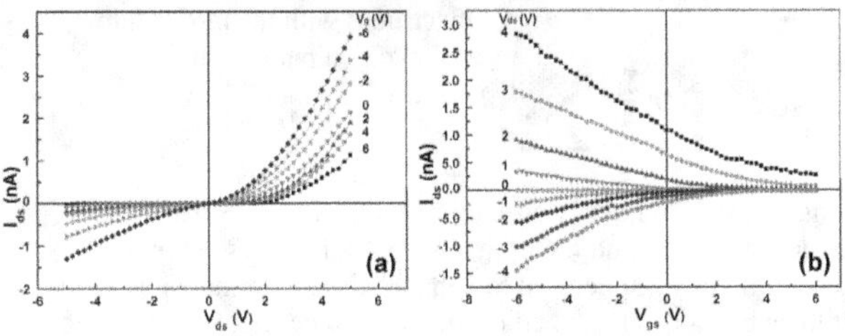

Figure 6 Electrical transportation characteristics of a typical device containing multiple CNT-bridges: (a) I_{ds}–V_{ds} curves and (b) I_{ds}–V_{g} curves.

In this study, it was found that laser polarization could significantly influence the selective growth of CNTs. To investigate the influence, cross-shaped electrode patterns were fabricated, as shown in Fig. 7(a). Each cross-shaped electrode was carved into two pairs of perpendicularly arranged tips, as shown in Figs. 7(b) and (c). For equal comparison and easy cutting, all electrode tips were fabricated with a tip angle close to 90°. A substrate temperature of 650 °C was used. The laser polarization was set to be parallel to one pair of the tips, as shown in Fig. 7(a). No external electrical field was used. CNTs were precisely wired between tips parallel to the laser polarization as shown in Figs. 7(b) and (c). Growth of CNTs showed obvious selectivity between the electrode tips parallel to the laser polarization. Again, a similar nonlinear I_{ds}–V_{ds} curve was observed at zero gate voltage, as shown in Fig. 7(d). The sharp RBM peak at 204.8 cm^{-1}, Fig. 7(e), indicated the growth of CNTs with a diameter of 1.21 nm [18-20]. The G^{-} band at 1565.5 cm^{-1} showed a Lorentzian lineshape, a characteristic of s-CNTs [21-23].

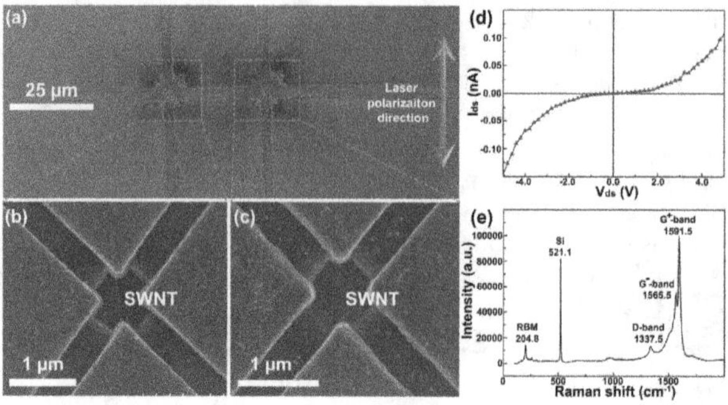

Figure 7 (a) An SEM micrograph showing the electrode pattern; (b) and (c) SEM micrographs of the squared regions in (a), showing the CNT- bridge structures; (d) *I*–*V* measurement of the device and (e) Raman spectrum of the CNTs.

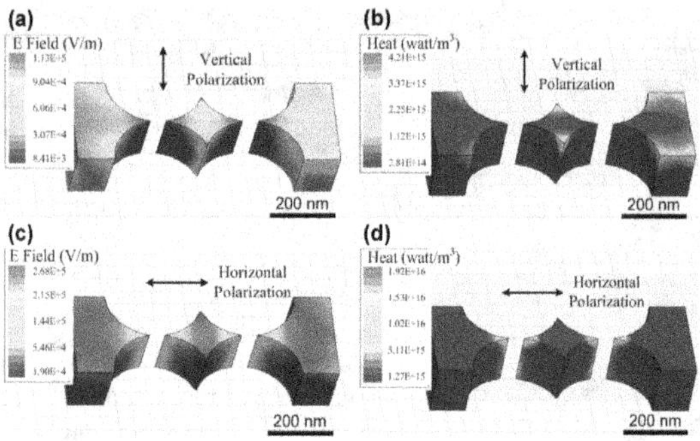

Figure 8 Electrical field distribution (a) and heat distribution (b) within the electrodes with the laser beam polarization normal to the electrode tip pair; electrical field distribution (c) and heat distribution (d) within the electrodes with the laser beam polarization parallel to the electrode tip pair.

Numerical simulations, using the high frequency structure simulator (HFSS, Ansoft), were carried out to investigate the mechanism behind the optically controlled integration of CNTs. It is well known that sharp metallic tips can be used to enhance the electromagnetic field at nanoscale tip regions and result in an enhanced localized heating at the apexes of the metallic tips [24,25]. In the simulation model, the incident laser beam was assumed to be linearly polarized and propagating perpendicularly to the substrate surface. The electromagnetic frequency was 28.3 THz, equivalent to the CO_2 laser wavelength (10.6 μm). The tip-gap was assumed to be 200 nm. The tip radius was assumed to be 20 nm. The electrode material was Ru. Laser beams of different polarization directions were used separately in the simulation, perpendicular and parallel to the electrode alignment. Figures 8(a) and (c) show the electrical field distribution in the metallic tips under laser irradiation of different polarization directions. Significant enhancement of the local electrical field is observed at the tips parallel to the laser polarization, almost tenfold stronger than that of the rest of the electrodes. Since a changing electromagnetic field generates an eddy current, the enhanced high-frequency E-field at the tips results in a significantly enhanced local eddy current and yields an enhanced localized heating at the nanoscale. Figures 8(b) and (d) show the heat distributions in the metallic tips under laser irradiation of different polarization directions. The local heating enhancement reaches the maximum at the tips parallel to the laser polarization, almost one order of magnitude larger than that of the rest of the electrodes. Therefore, a significant temperature increase at the tips is induced by the optical near-field effects. The simulation results clearly show that both optical near-field and localized heating reach their maximum values at the electrode tips in parallel with the laser polarization. Therefore, the highest temperatures are found at the electrode tips parallel to the laser polarization. A sufficiently high temperature will stimulate the growth of CNTs. Therefore, CNTs were observed bridging electrode tips parallel to the laser polarization, as shown in Figs. 7(b) and (c). For electrode tips perpendicular to the laser polarization, no obvious change was observed. Therefore, the growth position and orientation control was ascribed to the selective heating of electrode tips caused by the polarization dependent optical near-field effect.

2. Selective removal of metallic CNTs in open air

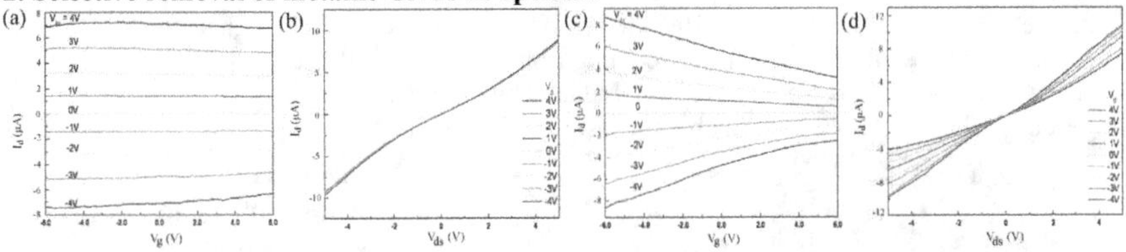

Figure 9 I_d–V_g and I_d–V_{ds} curves of the original sample before, (a) and (b), and after, (c) and (d), laser irradiation, with a laser wavelength of 2 μm and irradiation duration of 20 s.

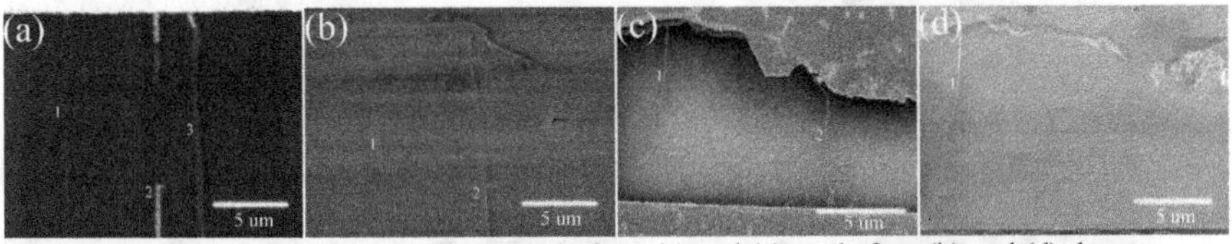

Figure 10 SEM images of the CNT sample before, (a) and (c), and after, (b) and (d), laser irradiation. Each number marks to the same CNT before and after laser irradiation.

In situ selective removal of m-CNTs was achieved through a laser irradiation process. Well aligned CNT arrays between metallic electrodes were fabricated. Electrical transport measurements were carried out before and after laser irradiation. CNT arrays containing m-CNTs exhibited linear electric transport features and high current values independent of the gate voltage. Figures 9(a) and (b) show the I_d–V_g and I_d–V_{ds} curves of a CNT array before laser irradiation with obvious existence of m-CNTs confirmed by the linear curves, high constant current values, and absence of V_g dependence. The CNT arrays then were irradiated using pulsed laser beams with a wavelength of 2 μm and a threshold power of 1.8 MW/mm^2 per pulse for 20 s. The threshold laser fluence was decided by increasing the diameter of spot size for a given laser intensity and observing the changes. I–V curves were measured again after the laser irradiation using the same experimental setup. As shown in Figs. 9(c) and (d), I–V curves show nonlinear semiconducting behavior, obvious V_g dependence, and significantly reduced current. SEM images of the same sample were taken before and after the laser irradiation to compare the physical changes of the CNTs, as shown in Fig. 10. All CNTs were labeled before laser irradiation. After laser irradiation, CNTs 3 and 2 are removed in Figs. 10(b) and (d), respectively.

Raman spectroscopic investigations were carried out to verify the selective removal of the m-CNTs. As reported, the tangential modes of s-CNTs contain only Lorentzian components for both G^- and G^+ frequency modes, while m-CNTs exhibit two strong peaks with the higher frequency mode G^+ having a Lorentzian line shape and the lower frequency mode G^- exhibiting an asymmetric Berit–Wigner–Fano (BWF) line shape [21,22,26]. Figure 11(e) shows the Raman spectra from three different regions on the sample, corresponding to the SEM images, starting from the OPO laser spot centre, Fig. 11(d), and regions away from the centre, Figs, 11 (b) and (c). In the OPO laser spot, the G-band shows Lorentzian line shape only, indicating the existence of s-CNTs only in this region. In the region away from the OPO laser spot, the G-band turns to showing a BWF line shape, indicating the existence of m-CNTs. According to the Kataura plot [23], the RBMs between 225 and 290 cm^{-1} come from m-CNTs due to their resonance with the 488 nm laser beam. The insets in Fig. 11(e) show zoomed views of the RBM signals from different areas of the sample. RBM signals coming from m-CNTs (in the rectangular boxes) were observed in the regions which were not irradiated by the laser beams. Only RBM peaks arising from s-CNTs were observed within the laser irradiated region.

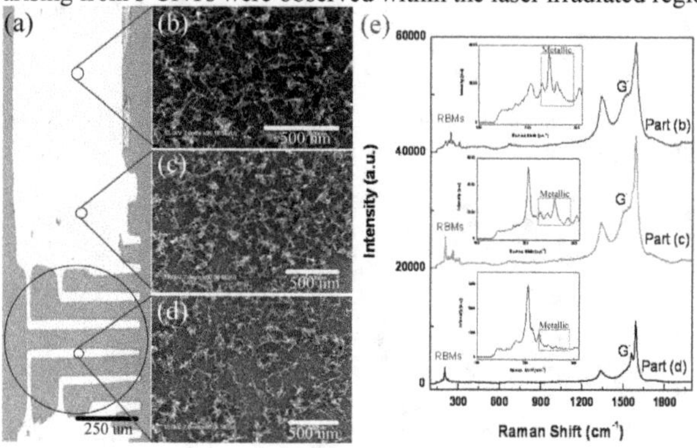

Figure 11 (a) An optical image showing an electrode with a circle indicating the OPO laser spot. (b), (c) and (d) SEM images collected at the indicated regions. (e) Raman spectra obtained from the selected regions corresponding to the SEM pictures, the insets are the close-up views of the RBM peaks.

Two major mechanisms were suggested to explain the selective removal of the m-CNTs through the optically controlled approach. The first is the free electron movement within the m-CNTs, which respond quickly to the electromagnetic field oscillation caused by the laser beam. However, s-CNTs are subject to change only when the photon energy is equal to or larger than their band energy. The second is the optical near-field effects arising in the m-CNTs, which would cause significantly enhanced eddy current and Joule heating in the m-CNTs. Both mechanisms contribute to the selective heating of m-CNTs. Therefore preferential heating and oxidation of m-CNTs are induced when the process is carried out in an open atmosphere.

3. Growing CNTs of controlled-alignments

CNT arrays of controlled alignments were grown on metallic electrodes. As shown in Fig. 12(a), surface-bounded CNTs were observed on the anode sticking out from the electrode and crawling on the SiO_2 surface. A fuzzy edge was observed. Vertically aligned CNTs were observed on the cathode, which showed a neat and clean edge. Raman spectra were measured on the anode and cathode, as shown in Figs. 12(b) and (c), respectively. A stronger RBM intensity was observed on the anode, while the cathode shows stronger D- and G-bands. The difference in the Raman spectra was ascribed to the alignments of the CNTs. The surface crawling CNTs have large Raman scattering cross-sections. Therefore, stronger RBM signals were observed in Fig. 12(b). Vertically aligned CNTs (with end caps exposing to the surface) have small Raman scattering cross-sections which lead to weak RBM peaks. The stronger D- and G-bands indicate more carbon deposition, as shown in Fig. 12(c). To confirm the surface-bounded growth of CNTs on anodes, E-fields of different intensities were applied, as shown in Figs. 13(a) to (d). Fuzzy edges with CNTs sticking out were observed on the anodes. The CNTs grown on the anodes were observed to stretch out from the electrodes and align with the E-fields. For the same growth period, an increased CNT length was observed under a stronger E-field.

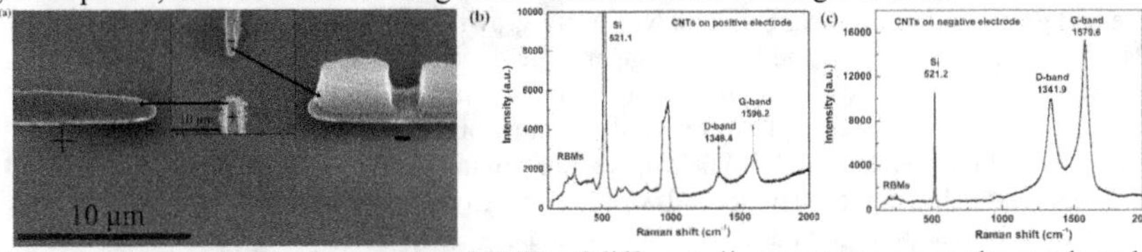

Figure 12 Typical SEM micrographs of CNTs of different alignments grown on the anode and cathode (a), and typical Raman spectra of the CNTs grown on the anode (b) and cathode (c.

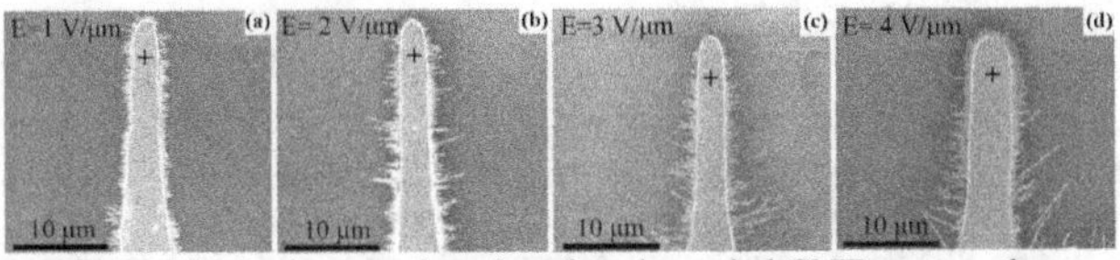

Figure 13 (a) to (d) are SEM micrographs of surface-bounded CNTs on anodes grown under different electric fields, 1, 2, 3, and 4 V/μm, separately.

A cross-shaped electrode pattern was used to study the growth of CNTs with different alignments influenced by the E-field distributions, as shown in Figs. 14(a) to (c). The electrodes were numbered in a clockwise order starting from the top electrode. In Fig. 14(a), fuzzy edges with crawling CNTs were observed on the anodes 3 and 4. Some CNTs were observed to stretch out from the cathode 2, which is ascribed to the formation of the CNT-bridges between the electrodes 2 and 3. Formation of the conductive CNT-bridge between the electrodes 2 and 3 made the two electrodes equipotential, which resulted in the observation of randomly distributed CNTs. In Fig. 14(b), surface-bounded CNTs were observed only on the anodes 3. A very clean edge was observed for the cathode 1. Randomly distributed CNTs were observed on the electrodes 2 and 4. In Fig. 14(c), long CNTs were observed between the electrodes 1 and 2. It seems that growth of CNTs on the anode 3 was suppressed compared with the electrodes 2 and 4 due to the weaker E-field between the electrode 3 and electrodes 2 and 4.

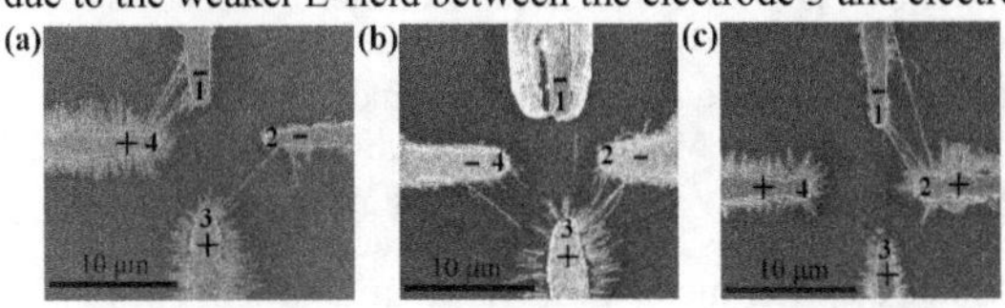

Figure 14 (a) to (c) are SEM micrographs of CNTs grown on electrodes under different E-field distributions

From above results, it can be concluded that CNT alignments can be controlled by biasing electrodes with different polarities. Anodes yield surface-bounded CNTs, while cathodes yield vertically aligned CNTs. CNTs stretch out from the anodes toward the cathodes following the E-field. It was suggested that movement of catalyst nanoparticles (NPs) was responsible for the CNT alignment. As shown in the inset of the Fig. 15(a), bright catalyst NPs are observed at the ends of the CNTs. In Fig. 15(b), a shiny top layer was observed, which indicates the coalesced catalyst NPs. Therefore, a tip-growth mode is suggested for the CNT growth in the LCVD process. The Fe NPs are positively charged due to the stronger electron affinity of carbon (Fe of 4.67 eV and CNT of 5.44 eV) [27]. As a result, the catalyst NPs are repulsed by the anodes and move away from the anodes toward the cathodes following the E-field. Since CNTs nucleate and grow from carbon-saturated Fe NPs, CNT growth follows the moving tracks of the Fe NPs, as shown in Figs. 15(c) and (d). A stronger E-field would exert stronger force on the Fe NPs and yield longer CNTs, which supports the observation in Fig. 13. For the CNTs grown on the cathodes, vertically aligned CNTs were ascribed to the joint contribution from following interactions. The first was the van der Waals interaction among the CNTs, which made CNTs entangled and vertically aligned. The second was the attractive force between the positively charged catalyst NPs and the cathodes, which prevented CNTs from stretching outside. The third was the repulsing forces caused by the continuous growth of CNTs, which kept pushing the catalyst NPs outward. Therefore, the Fe NPs were forced to move outside of the surface plane and result vertically aligned growth of CNTs.

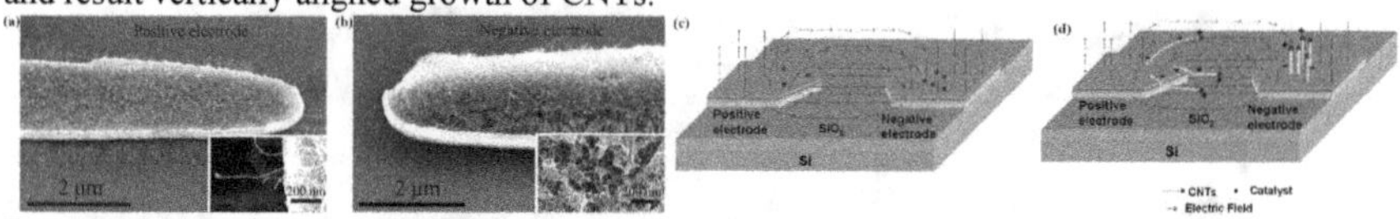

Figure 15 (a) and (b) are SEM micrographs of CNTs on the anode and cathode, respectively. (c) and (d) are schematic diagrams of CNT growth induced by the movement of Fe NPs in E-field.

4. Diameter modulation in individual CNTs

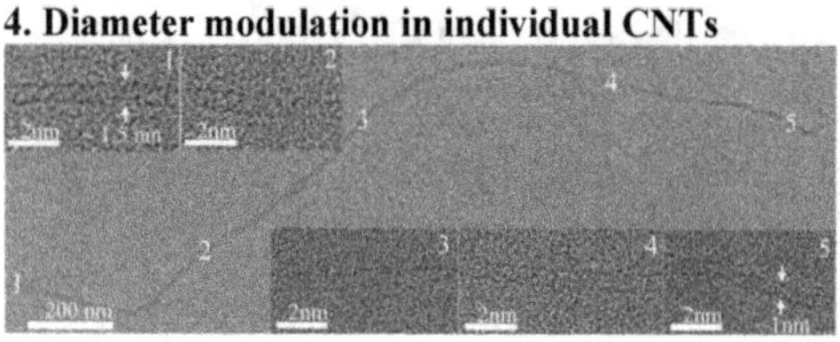

Figure 16 TEM micrographs of a CNT grown with a descending temperature profile. The close up views, 1–5, correspond to the labeled areas of the tube, show 0.5 nm variation in the diameter from one end to the other.

Diameter-modulated CNTs were obtained by adjusting reaction temperatures during the LCVD growth process. A descending temperature profile from 650 to 450 °C was adopted to grow single-walled CNTs (SWNTs) with increasing diameters along their axial directions. Figure 16 shows the TEM images of a typical SWNT with an increasing diameter from one end to the other. Electronic transport properties of the diameter-modulated SWNTs were investigated by fabricating a CNT-FET structure using the diameter-modulated SWNTs. Individual diameter-modulated SWNTs were grown across the Mo electrodes. Micro-Raman spectroscopy was conducted to map along their axial directions. In addition, I–V characteristics were measured to determine the electronic transport characteristics of the CNT-FETs. I–V characteristics and Raman spectra of the devices fabricated with a diameter-modulated SWNT are shown in Figs. 17. The device with a diameter-modulated SWNT has a diode-like I–V curve. The device with the diameter-modulated SWNT forms two different contacts with the metallic electrodes. One side forms a Schottky barrier between the Mo electrode and the SWNT end with a smaller diameter (having wider bandgap). The other forms an Ohmic contact between the Mo electrode and the SWNT end with a larger diameter (smaller bandgap). The band diagram of the device is shown in the inset of Fig. 17(d). When the device was positively biased ($V_{ds} > 0$) and the Schottky contact (source) was grounded, a positive current value was obtained. When negatively biased ($V_{ds} < 0$), no current flowed through the device. This approach significantly simplifies the processes to fabricate SWNT-based Schottky diodes. Micro-Raman spectroscopy of the SWNTs was conducted by mapping the SWNTs in 1 μm steps to investigate the diameter modulation and corresponding bandgap change along the SWNTs, as assigned in the Kataura Plot [23]. As shown in Fig. 17(b), the RBM peak shifts from a high frequency (high-temperature end) to a low frequency (low-temperature end) of the SWNT. The double-peak feature obtained at position 2 of the SWNT indicates an inter-junction area of the SWNT since the higher peak is the same as that of position 1 and the lower peak is the same as that of position 3 of the SWNT. It is observed that the electrical currents are very low in diameter-modulated SWNTs due to electron scattering caused by the defects created by the diameter modulation.

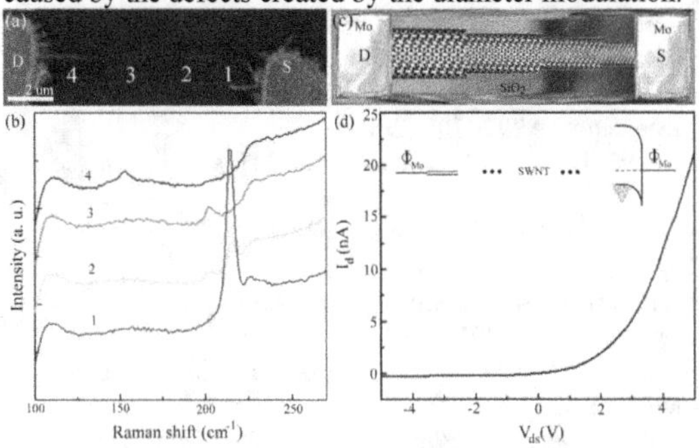

Figure 17 (a) A SEM image of a single SWNT bridging a pair of Mo electrodes; (b) Raman spectra of different sections of the SWNT in (a); (c) the schematic representation of the SWNT structure; and (d) typical *I–V* characteristics of the device and an expected band diagram of the CNT.

CONCLUSIONS

In this study, laser-based strategies were developed to achieve controlled growth, removal and integration of CNTs. Parallel integration of CNTs into micro/nano-architectures was successfully achieved at a relatively low substrate temperature by integrating optical near-field effect into a LCVD process. *In situ* selective removal of m-CNTs was realized through an optically controlled approach. CNT arrays of controlled alignments were obtained by applying voltages of different polarities. Diameter modulated CNTs were successfully fabricated through precise temperature tuning in the LCVD growth process. The laser-based strategies provide new insight into the feasible and scalable nano-fabrication of CNT-based devices and stimulate successful implementation of the CNTs.

ACKNOWLEDGMENTS

The authors would like to express thankfulness to the National Science Foundation (ECCS 0621899 and ECCS 0652905) and Nebraska Research Initiative for their financial support. The authors are grateful to Drs You Zhou, Natale Joseph Ianno and Lan Ping Yue from the University of Nebraska-Lincoln (UNL) for their technical supports. We would also like to express our gratefulness to Joel Brehm from Office of Research, UNL, for his graphic designs.

REFERENCES

1. P. Avouris, R. Martel, V. Derycke and J. Appenzeller, *Physica B* **323**, 6 (2002).
2. M. Burghard, H. Klauk and K. Kern, *Advanced Materials* **21**, 2586 (2009).
3. A. Bachtold, P. Hadley, T. Nakanishi and C. Dekker, *Science* **294,** 1317 (2001).
4. H. J. Dai, *Surface Science* **500**, 218 (2002).
5. P. Avouris and J. Chen, *Materials Today* **9**, 46 (2006).
6. N. R. Franklin, Q. Wang, T.W. Tombler, A. Javey, M. Shim and H. J. Dai, *Applied Physics Letters* **81**, 913 (2002).
7. M. R. Falvo, G. J. Clary, R. M. Taylor, V. Chi, F. P. Brooks, S. Washburn and R. Superfine, *Nature* **389**, 582 (1997).
8. Y. Y. Wei and G. Eres, *Nanotechnology* **11,** 61 (2000).
9. K. Keren, R. S. Berman, E. Buchstab, U. Sivan and E. Braun, *Science* **302**, 1380 (2003).
10. J. Q. Li, Q. Zhang, N. Peng and Q. Zhu, *Applied Physics Letters* **86**, 153116 (2005).
11. R. Krupke, F. Hennrich, V. H. Lohneysen and M. M. Kappes, *Science* **301**, 344 (2003).
12. P. G. Collins, M. S. Arnold and P. Avouris, *Science* **292**, 706 (2001).
13. S. Fan, M. G. Chapline, N. R. Franklin, T. W. Tombler, A. M. Cassell and H. J. Dai, *Science* **283**, 512 (1999).
14. Y. T. Liu, X. M. Xie, Y. F. Gao, Q. P. Feng, L. R. Guo, X. H. Wang and X. Y. Ye, *Materials Letters* **61**, 334 (2007).
15. S. M. Huang, M. Woodson, R. Smalley and J. Liu, *Nano Letters* **4**, 1025 (2004).
16. Y. Avigal and R. Kalish, *Applied Physics Letters* **78**, 2291 (2001).
17. Y. Nakayama, L. J. Pan and G. Takeda, *Japanese Journal of Applied Physics Part 1-Regular Papers Brief Communications & Review Paper* **45**, 369 (2006).
18. M. S. Dresselhaus, G. Dresselhaus, R. Saito and A. Jorio, *Physics Reports-Review Section of Physics Letters* **409**, 47 (2005).
19. A. G. Souza, S. G. Chou, G. G. Samsonidze, G. Dresselhaus, M. S. Dresselhaus, L. An, J. Liu, A. K. Swan, M. S. Unlu, B. B. Goldberg, A. Jorio, A. Gruneis and R. Saito, *Physical Review B* **69,** 115428 (2004).
20. A. Jorio, R. Saito, J. H. Hafner, C. M. Lieber, M. Hunter, T. McClure, G. Dresselhaus and M. S. Dresselhaus, *Physical Review Letters* **86,** 1118 (2001).
21. R. Saito, A. Jorio, J. H. Hafner, C. M. Lieber, M. Hunter, T. McClure, G. Dresselhaus and M. S. Dresselhaus, *Physical Review B* **64,** 085312 (2001).
22. S. D. M. Brown, A. Jorio and P. Corio, *Physical Review B* **63**, 155414 (2001).
23. H. Kataura, Y. Kumazawa, Y. Maniwa, I. Umezu, S. Suzuki, Y. Ohtsuka and Y. Achiba, *Synthetic Metals* **103**, 2555 (1999).
24. N. Hayazawa, T. Yano, H. Watanabe, Y. Inouye and S. Kawata, *Chemical Physics Letters* **376,** 174 (2003).
25. L. Novotny, R. X. Bian and X. S. Xie, *Physical Review Letters* **79**, 645 (1997).
26. A. Jorio, A. G. Souza Filho, G. Dresselhaus, M. S. Dresselhaus, A. K. Swan, M. S. Ünlü, B. B. Goldberg, M. A. Pimenta, J. H. Hafner, C. M. Lieber and R. Saito, *Physical Review B* **65**, 155412 (2002).
27. G. Zhou, W. Duan and B. Gu, *Physics Review Letters* **87**, 095504 (2001).

Mater. Res. Soc. Symp. Proc. Vol. 1365 © 2011 Materials Research Society
DOI: 10.1557/opl.2011.1319

Micro-Structuring of Metal Films by Localized Single-Pulse Laser Irradiation

Chandra Prakash R. Gaddam[1], Joseph P. Moening[2], Nanke Jiang[1], Daniel G. Georgiev[1]

[1] Department of Electrical Engineering and Computer Science, University of Toledo, Toledo, OH 43606, U.S.A.

[2] Department of Electrical Engineering, Lake Superior State University, Sault Ste. Marie, MI 49783, U.S.A.

ABSTRACT

We present data on micrometer-scale localized single-pulse laser irradiation of Au, Cu, Al, or Ti films on borosilicate glass substrates. These metals represent a range of thermal properties, chemical reactivity levels, and relevance to specific applications. A mask projection technique employing a Q-switched Nd:YAG laser, emitting at its fourth harmonic of 266nm, was used to produce the irradiation spots in this work. The metal films, deposited by RF-sputtering, had thicknesses of several hundred nanometers. Sample irradiation was performed in either vacuum or ambient air, and the resulting microstructures were examined by electron microscopy. The results indicate that irradiation of Cu films can lead to the formation of bumps, sharp cones or protrusions. However, the controllability of these structures on Cu films is limited, compared to those formed on Au or Si. The results, upon irradiation of Ti films, are limited to melting and surface roughening or ablation openings, regardless of the conditions of irradiation, film thickness, substrate or ambient gas. The modifications that occur within Al films are reproducible, but limited in shape and size.

INTRODUCTION

Pulsed laser radiation can be used to modify surfaces of materials and fabricate structures on a micrometer and sub-micrometer scale [1-4]. In addition to their technological importance, such surface modifications are of general scientific interest as well. The mechanisms of formation of such structures are not well understood due to their complexity, and therefore, systematic studies of laser irradiation of such films as a function of various parameters are essential. We have reported the conditions for controllable, direct laser fabrication of sharp conical nano-tips on silicon thin films [4] as well as sharp high-aspect ratio protrusions or microbumps on gold films [3]. Such laser-based approaches *promise to provide reliable, simple and low-cost methods for the fabrication of micro- and nano-scale structures of semiconductor and metal materials.* Potential applications include emitters for field-emission-based devices such as high-definition displays [5,6] and other vacuum microelectronics applications [7], surface patterning of materials for biomedical[8,9], and catalyst applications, as well as micro and nano-patterning steps for microelectronics and sensor applications in general.

In this work, *which is motivated by both the potential applications and the goal of understanding the mechanism of formation*, we present new results on localized pulsed-laser irradiation of metal films. These include an extension of our work on gold films, as well as our first results from the irradiation of copper, aluminum and titanium films. Representative cases from several specific combinations of film thickness, substrate, spot size, and fluence level are shown and briefly discussed.

EXPERIMENTAL DETAILS

Films of copper, aluminum or titanium of various thicknesses were deposited on borosilicate glass (BSG) or single-crystal silicon substrates by a RF magnetron sputtering technique using a Torr International deposition system, operating at a base vacuum of 10^{-6}Torr. Gold films were deposited separately in a DC sputtering machine and a very thin Ti film was pre-deposited to serve as an adhesion-promoting layer.

A Q-switched Nd:YAG laser (Powerlite 8010 from Continuum Laser, Inc), delivering energy of up to 200mJ in 8ns pulses at a wavelength of 266nm (4th harmonic) was the laser source used in this work. The laser irradiation set-up employed a diffraction-limited lens in a projection geometry, serving to image individual, or small arrays of pinholes onto the gold film. A demagnification factor of 10 was used in order to produce circular spots of almost uniform intensity. The irradiation was performed under ambient or low-vacuum conditions, by using a compact vacuum/gas chamber, with a top window of UV-grade fused silica. Single-pulse laser irradiation with fluence levels between 0.05 – 1.5 J/cm^2 was used on each of the sample films.

The resulting structures were examined by optical microscopy and then imaged at higher magnification by scanning electron microscopy (SEM), on a Hitachi S-4800 UHR scanning electron microscope.

RESULTS AND DISCUSSION

Irradiation of Au films

In an earlier publication [3], we reported the formation of various types of structures as a result of single-pulse irradiation of gold films on glass substrates with circular spots several micrometers in diameter. Different formation regimes with very different stability and reproducibility were identified, depending on the laser fluence. Figures 1 and 2 show new SEM images of the structures that form upon irradiation of 300nm-thick gold films, using 2.5μm-spots over wide range of fluence levels. The microbumps and small cones that form at low fluence levels (0.20 and 0.30 J/cm^2, Fig. 1a and b) are almost perfectly reproducible (with reproducibility better than 95%). Upon increasing the fluence, the height of the structure increases significantly with a drop beginning to form on top. The fluence range corresponding to this case is 0.4 – 0.6 J/cm^2 (Fig. 1c-f). The shape and size of the resulting structures is rather random and much less reproducible. In addition, some of the droplets on top get detached from the central shaft. The fluence range 0.6 – 1.0 J/cm^2 (Fig.2a-d), corresponds to the reproducible formation (with reproducibility better than 90%) of sharp high-aspect ratio cones or protrusions. The heights of these protrusions can be controlled (within a substantial range) by the laser fluence: where increasing the fluence leads to a decrease in the height. The formation of these structures apparently involves a pinching-off and detachment of a droplet from the top of a jet of molten gold, which then solidifies in this shape [3]. We have also found that the underlying bump, which increases in size with the fluence, is hollow, whereas the central shaft is solid [10]. Further increases in the fluence leads to the formation of even larger underlying bumps (Fig. 2e and f), but with less control over their shape, and eventually to the formation of ablation holes.

We note that the fluence formation range for the various structures, depends on the combination of spot size and film thickness. Most generally, thicker films and smaller spot sizes

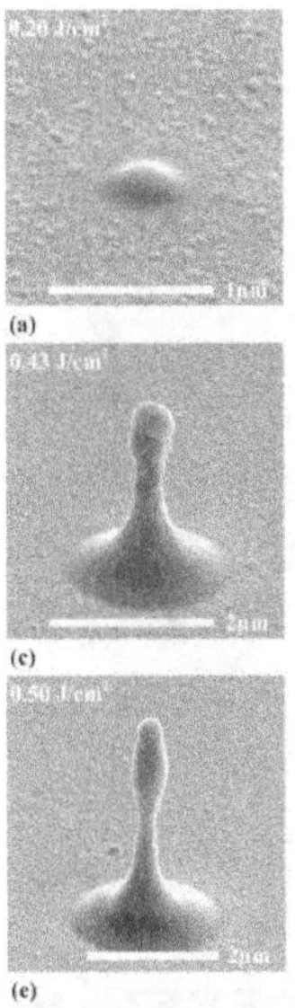

Figure 1. SEM images of structures resulting from single-pulse irradiation of 300nm-thick gold films on glass substrates with circular spots, 2.5μm, in diameter in low vacuum, at the indicated fluence levels (lower fluence).

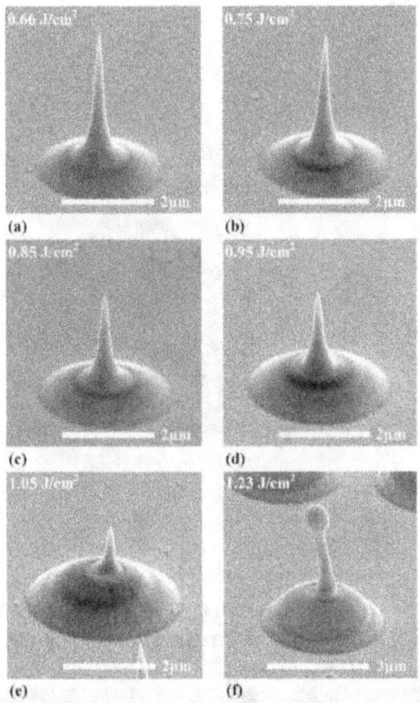

Figure 2. SEM images of structures resulting from single-pulse irradiation of 300nm-thick gold films on glass substrates with circular spots, 2.5μm, in diameter in low vacuum, at the indicated fluence levels (higher fluence).

require higher fluence levels. There is an upper limit to the spot size, beyond which these sharp structures do not form; instead, one obtains a roughened surface or ablation holes. The substrate type (especially its thermal conductivity) plays a role in the formation of these structures.

<u>Irradiation of Cu films</u>

We were also interested in studying the case of irradiation of copper films under conditions similar to those of the gold films as described above. Copper is similar to gold in terms melting point, but has slightly better thermal conductivity. Given the similarities in thermal properties, there was an expectation that comparable types of structures could form within copper films. This would significantly expand the technological interest in this type of fabrication. Copper is much more chemically reactive than gold, which may be of importance to the formation process, together with the effect of the morphology of the films. Shown below are SEM images of the surface modifications that resulted from single-pulse irradiation with 2.5μm and 5μm diameter spots under low vacuum conditions (to avoid any oxidation effects) for two different film thicknesses. In Fig. 3, one can see that at relatively low fluence, reproducible formation (with reproducibility better than 90%) of microbumps and small protrusions occurs upon irradiation of 250nm-thick films, which is very similar to the lower-fluence case seen in the irradiation of gold films.

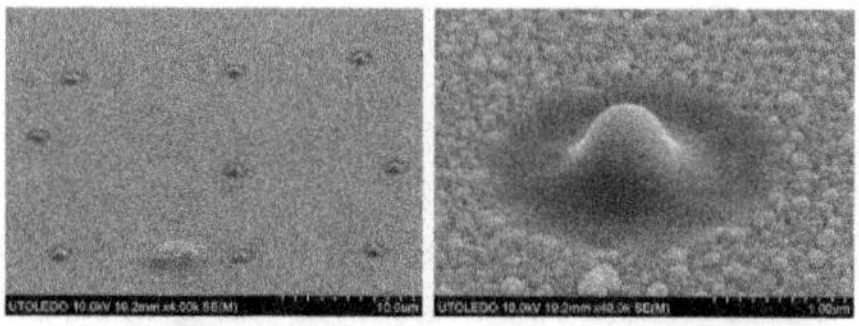

(a) 250nm Cu on BSG, 0.4J/cm^2 250nm Cu on BSG, 0.4J/cm^2

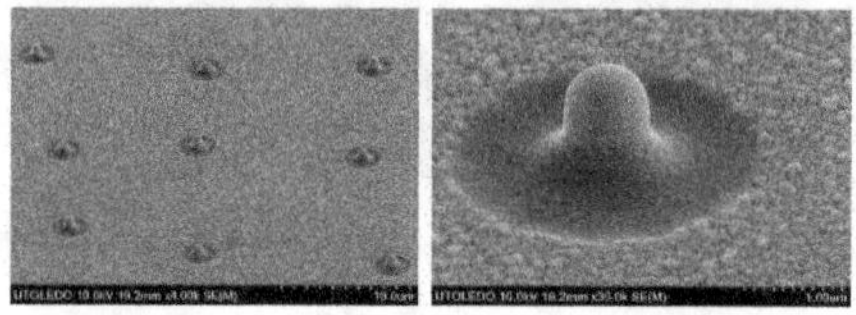

(b) 250nm Cu on BSG, 0.8J/cm2 250nm Cu on BSG, 0.8J/cm2

Figure 3. SEM images of structures resulting from single-pulse irradiation of 250nm-thick Cu films on glass substrates with circular spots, 2.5 μm, in diameter in low vacuum, at the indicated fluence levels

In Fig.4 and 5 are shown results from the irradiation of 500nm-thick films at similar fluence levels, which have generally resulted in somewhat sharper, taller and less reproducible

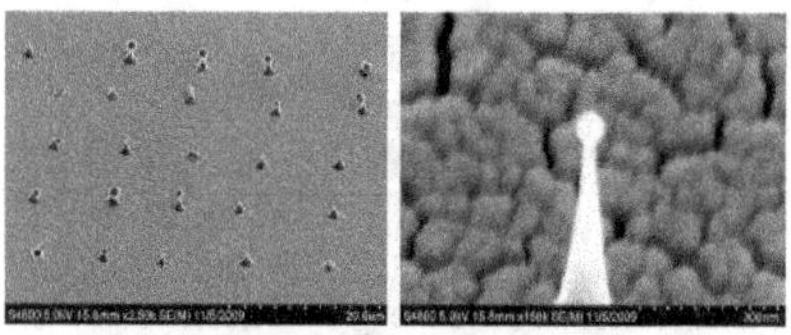

(a) 500nm Cu on BSG, 0.45J/cm^2 500nm Cu on BSG, 0.45J/cm^2 (close-up)

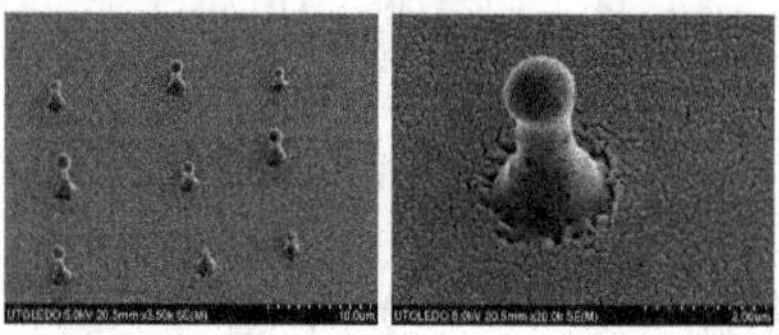

(b) 500nm Cu on BSG, 0.7J/cm^2 500nm Cu on BSG, 0.7J/cm^2 (close-up)

Figure 4. SEM images of structures resulting from single-pulse irradiation of 500nm-thick Cu films on glass substrates with circular spots, 2.5μm, in diameter, in low vacuum, at the indicated fluence levels. Note that the film surface contains large density of cracks.

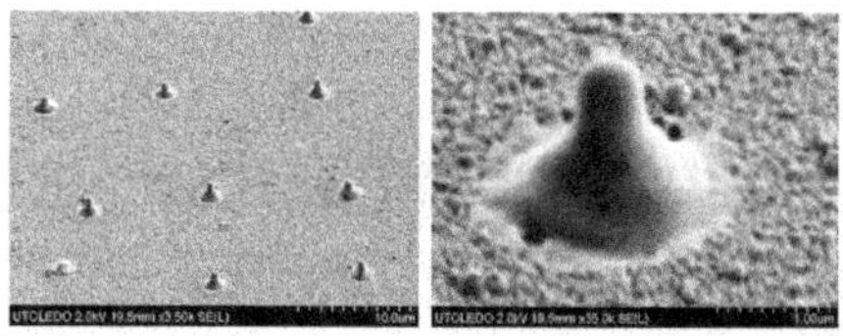

(a) 500nm Cu on BSG, 0.8J/cm^2 500nm Cu on BSG, 0.8J/cm^2 (close-up)

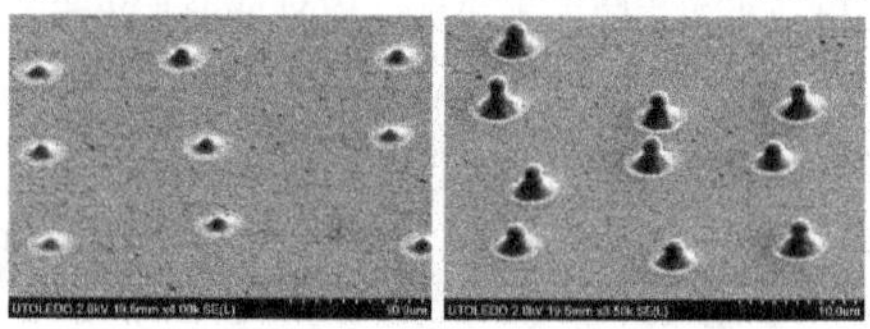

(b) 500nm Cu on BSG, 0.4J/cm^2 (c) 500nm Cu on BSG, 0.6J/cm^2

Figure 5. SEM images of structures resulting from single-pulse irradiation of 500nm-thick Cu films on glass substrates with circular spots, 2.5μm (a) and 5μm (b and c), in diameter, in low vacuum, at the indicated fluence levels. Note the difference in the resulting structures and the fact that the film has no cracks (film was deposited at a substrate temperature of 200°C to improve its quality.

structures. The main reason for the differences appears to be related to the film surface morphology. It can be seen that the original surface of the film in Fig. 4, consists of small grains separated by significant voids or cracks, which can decrease the thermal conductivity and lead to stronger heat confinement within the irradiated spot. Almost no melting can be seen around the structures in Fig. 4, whereas the more dense film materials in Fig. 3 has been melted significantly around the structures, as evidenced by the smooth-surface depressions around perimeter.

Structures forming upon irradiation of a 500nm-thick copper film, which had a much more compact structure (i.e., no cracks), achieved by deposition at an elevated substrate temperature (200°C) are shown in Fig. 5. As is seen, the shapes obtained are much more reproducible, and clearly different from those in Fig. 4, with the melting depressions around the structures being clearly visible.

Irradiation of Al and Ti films

Representative SEM images of the results from irradiation of aluminum and titanium films on glass substrates are shown below in Figures 6 and 7. In the case of Al films, the general

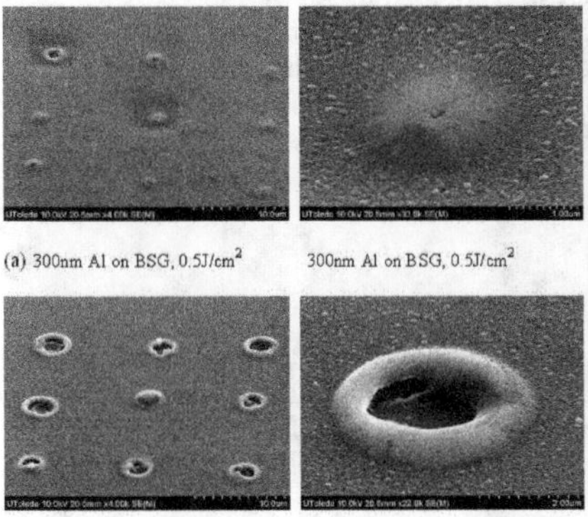

(a) 300nm Al on BSG, 0.5J/cm^2 300nm Al on BSG, 0.5J/cm^2

(b) 300nm Al on BSG, 0.9J/cm^2 300nm Al on BSG, 0.9J/cm^2

Figure 6. SEM images of structures resulting from single-pulse irradiation of 300nm-thick Al films on glass substrates with circular spots, 2.5μm, in diameter, in low vacuum, at the indicated fluence levels.

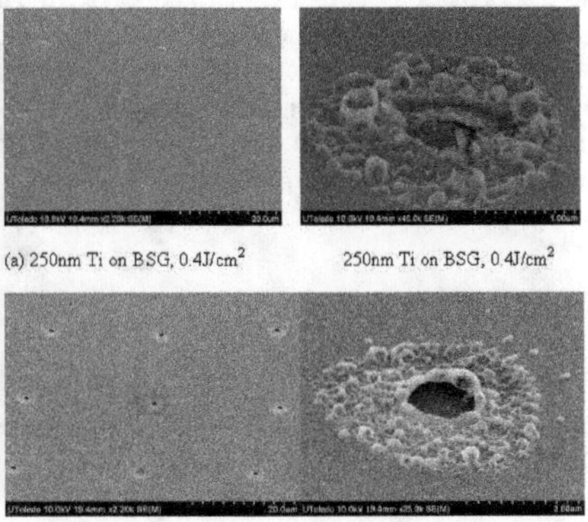

(a) 250nm Ti on BSG, 0.4J/cm^2 250nm Ti on BSG, 0.4J/cm^2

(b) 250nm Ti on BSG, 0.9J/cm^2 250nm Ti on BSG, 0.9J/cm^2

Figure 7. SEM images of structures resulting from single-pulse irradiation of 250nm-thick Ti films on glass substrates with circular spots, 2.5μm in diameter, in low vacuum, at the indicated fluence levels.

fluence dependence of the film modifications is as follows. At low and intermediate fluence levels, small, volcano-like microbumps form, with a central hole (Fig. 6a). Upon increasing the fluence, the bumps grow slightly in size, and the central hole opens wider. Even higher fluence levels cause the bumps to turn into ablation holes with raised edges as can be seen in Fig. 6b. This fluence dependence is representative for other combinations of film thickness and spot size, as well as for irradiation in ambient air environments. The resulting microbumps, or raised-edge-hole structures (that have a relatively low aspect ratio) are very reproducible in size and shape, but the sizes themselves can only be controlled over a narrow range.

In the case of Ti films (Fig. 7), we have not been able to obtain microstructure formation. The micrometer-scale localized irradiation results in a roughening of the surface within the laser-irradiated spot, and, at higher fluence levels, the formation of rough and uncontrollably shaped ablation holes. These observations apply to several spot size, fluence level, and film thickness combinations, on both glass and silicon substrates under both low-vacuum and ambient air conditions. The main reason for the different behavior of the Ti films appears related to its relatively low thermal conductivity, combined with its higher melting point. It is thought that this leads to significant local overheating and melting (i.e., melting and overheating at numerous hot spots).

CONCLUSIONS

We have studied the formation of micro-structures on thin metal films as a result of single-pulse micrometer-scale laser irradiation. Our results have shown that irradiation of Cu films can lead to the formation of bumps, sharp cones or protrusions. However, these structures have a limited controllability when compared to the formation of similar structures on Au films. The formation of such structures on Ti films appears limited to random melting and surface roughening or ablation openings, regardless of the conditions of irradiation, film thickness, substrate or ambient gas. The modifications of Al films on the other hand are reproducible, although limited in terms of the shapes and sizes that are achievable.

ACKNOWLEDGMENTS

The SEM imaging was done at the University of Toledo's Center for Materials and Sensor Characterization. The gold films used in this work were provided by Dr. S. Thanawala, from Greatbatch Inc., Clarence, NY.

REFERENCES

1. N.M. Bulgakova, A.N. Panchenko, A.E. Tel'minov, M.A. Shulepov, *Appl.Phys.* **A98**, 393 (2010).
2. D.G. Georgiev, R.J. Baird, I. Avrutsky, G. Auner, G. Newaz, Appl.Phys.Lett. **84**, 4881 (2004).
3. J.P. Moening, S.S. Thanawala, and D.G. Georgiev, *Appl.Phys.* **A95**, 635 (2009).
4. J.P. Moening, and D.G. Georgiev, *J. Appl. Phys.* **107,** 014307 (2010).
5. C.A. Spindt, I. Brodie, L. Humphrey, E.R. Westerberg, J. Appl. Phys. **47**, 5248 (1976).
6. L. Dvorson, I. Kymissis, A.I. Akinwande, J.Vac.Sci.Technol. B, **21**, 486 (2003).
7. H. H. Busta, J. Micromech. Microeng. **2** , 43 (1992).
8. A.M.P. Turner, N. Dowel, S.W.P. Turner, L. Kam, M. Isaacson, J.N. Turner, H.G. Craighead, W. Shain, J. Biomedical Res. **51**, 430 (2000).
9. M.P. Maher, J. Pine, J. Wright, Y.C. Tai, J. Neuroscience Methods **87**, 45 (1999).
10. J.P. Moening, D.G. Georgiev, and J.G. Lawrence, *J. Appl. Phys.* **109**, 014304 (2011).

Mater. Res. Soc. Symp. Proc. Vol. 1365 © 2011 Materials Research Society
DOI: 10.1557/opl.2011.890

Assembly of carbon nanotube devices by tip-induced optical trapping

W. Xiong, Y.S. Zhou, M. Mitchell, J.B. Park, M. M. Samani, Y. Gao, and Y.F. Lu
Department of Electrical Engineering, University of Nebraska-Lincoln,
Lincoln, NE 68588-0511.

ABSTRACT

Fabrication of nanoscale devices by assembling individual carbon nanotubes (CNTs) remains challenging despite enormous effort made in this field. Fulfilling the promise of CNTs requires more efficient assembly techniques. In this study, we have developed an *in-situ* assembly method for precise and cost-effective integration of CNTs using a laser-assisted chemical vapor deposition (LCVD) process. Results show that CNTs can be trapped between sharp tip-shaped electrodes due to the optical gradient forces around the tip apexes generated by a CO_2 laser irradiation. This method enables the precise assembly of CNT-based field-effect transistors (FETs) and paves the way for the successful implementation of the CNT-based nanoelectronics.

INTRODUCTION

Carbon nanotubes (CNTs) are building blocks for various functional devices including carbon-based integrated circuits [1], sensors [2], solar cells [3], and nanoelectromechanical systems (NEMS) [4] due to their unique structures and exceptional properties. For all of those potential applications, achieving scalable and precise integration of individual CNTs into desired micro- and nanoarchitectures remains a key and challenging issue. So far, various methods using different driving forces for CNT assembly have been investigated, such as CNT assembly by mechanical force [5], electrostatic force [6], van der Waals force [7], and E-field-induced dielectrophoretic force [8,9], of which the dielectrophoretic assembly is regarded as the most effective technique due to its parallel working mode and effective placement control. However, this technique suffers from unavoidable contamination due to the solution treatment. Moreover, the requirement for a physical contact between tiny electrodes and a radio frequency (RF) source makes the technique difficult in creating relatively complex integrated systems, such as integrated circuits containing many isolated electrodes. Therefore, to fulfill the promise of CNTs in electronics, a scalable assembly method with a nanometric precision and a noncontact characteristic is highly desired.

Optical tweezers are known to have the capability of noncontact manipulation of various micro- and nano-objects with high precision. A theory of nanometric optical tweezers has been proposed by L. Novotny in 1997 using strongly enhanced optical near-fields around sharp metal tips under laser irradiation [10]. The tip-enhanced optical near-fields can generate enhanced optical gradient forces which could be used to trap nano-particles at the tip apexes, resulting in nanometric manipulation precision [10]. However, most research works using nanometric optical tweezers were carried out in a liquid environment. And there is no report on applying optical gradient forces for the assembly of CNT devices.

In this study, we used localized optical gradient forces induced by the tip-enhanced optical near-fields to integrate CNTs with tip-shaped electrodes in a laser-assisted CVD process. It is demonstrated that, position-controllable, rapid and scalable integration of CNTs between electrode tips can be realized by the tip-induce optical traps. A p-type CNT-based field-effect transistor (FET) has been fabricated. This all-optical assembly method opens a new door to the fabrication of CNT-based nanoscale electronics and sensor devices.

EXPERIMENT

CNTs were synthesized using a laser-assisted chemical vapour deposition (LCVD) method [11]. Figure 1 shows the schematic of the experimental setup. Heavily doped Si wafers covered with a 2 μm thick SiO_2 layer were used as the substrates. The electrode patterns were fabricated by conventional photolithography followed by DC sputtering. 200 nm thick Ruthenium (Ru) films and Al/Fe/Al tri-layer (15 nm/1.5 nm/5 nm) catalyst thin films were deposited sequentially on the SiO_2 surface. Ru was used as the electrode material due to its low solubility with Fe, high work function, and high melting point. Iron serves as the active catalyst material for growing carbon nanotubes. The Al layers beneath and above the iron layer were used as the buffer layers for restructuring the thin iron film into catalytically active islands on Si/SiO_2 substrate [12]. Sharp electrode tips were fabricated by focused ion beam (FIB) nano-machining using an FEI Strata 200xp system. The electrode gap distance was controlled to be around 500 nm. Growth of CNTs was carried out in an LCVD chamber. A continuous-wave (CW) CO_2 laser (Synrad, firestar v40, wavelength 10.6 μm, linear polarized, laser beam diameter 2.5 mm ± 0.5 mm) was used to irradiate the patterned substrates. The laser power was adjusted to maintain a stable substrate temperature. During the LCVD process, a gas mixture of acetylene (C_2H_2) and anhydrous ammonia (NH_3) with a volume ratio of 1:10 was introduced into the chamber. The C_2H_2 provides the carbon source under high temperature. The anhydrous ammonia is used as a buffer gas in the LCVD process to dilute acetylene, create an etching environment to suppress the growth of amorphous carbon, and protect catalyst particles from being poisoned by amorphous carbon [13]. The reaction pressure was maintained at 10 Torr. The reaction temperature for CNT growth was controlled around 500 °C, which was 200 °C lower than that required in the conventional LCVD process [11]. The whole LCVD process only took 3 min.

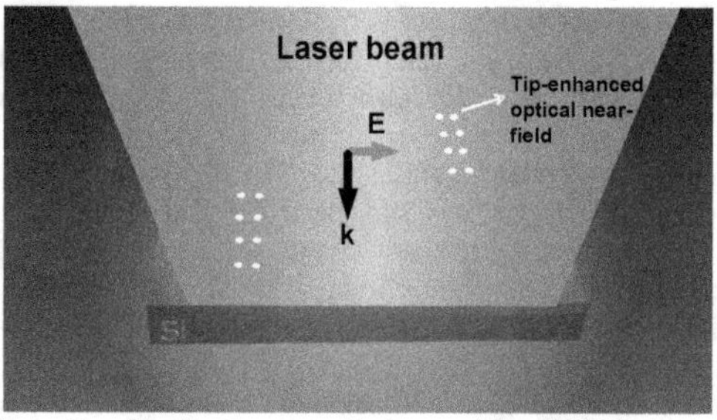

Figure 1. Schematic of optically controlled assembly of CNT devices by LCVD.

The CNT-based devices were characterized using a Hitachi S4700 field-emission scanning electron microscope (SEM) with an ultimate resolution of 1.2 nm at 25 kV. Electrical transport measurements were taken with an Agilent HP4155C semiconductor parameter analyzer with a 100 fA resolution. Raman spectra were performed on a Renishaw inVia Raman microscope with a 488 nm laser excitation source.

DISCUSSION

The SEM image of a CNT-based FET fabricated by the LCVD is shown in Figure 2. The device has a single CNT bridging two tip-shaped electrodes which function as source and drain terminals. The CNT bridge and the heavily-doped Si substrate function as a channel for hole transport and a backgate terminal, respectively. The CNT is precisely connected to the tip apexes with a sub-20-nm radius of tip curvature, suggesting the high accuracy of the optical assembly method. Moreover, looking closely at the CNT/tip contacts, as shown in Figure 2a, we found that although the CNT did not grow exactly from the tip apexes but rather from somewhere else in the vicinity of the tip apexes, the CNT always passed through the tip apexes. This behavior of the CNT indicated the existence of trapping forces towards the tip apexes. Our previous studies have shown that laser irradiation could generate localized heating at the tip areas due to the tip-enhanced optical near-fields which promote the site-selective growth of CNTs in the vicinity of the tip apexes [14,15]. However, the site-selective growth is not, by itself, sufficient to explain the orientation alignment and precise assembly of individual CNTs at the tip apexes. Attractive forces should also be involved in the LCVD process for the CNT assembly. It is noteworthy that no electric field was applied between electrode tips, thus the electrostatic driving force can be excluded. Therefore, the dominant assembly force is among the following candidates: gas flow, van der Waals force and optical gradient force. To further indentify the assembly force in the LCVD process, a control experiment was conducted with a backside heating configuration (the laser beam irradiated the backside of the Si substrate to heat up the electrode patterns on the front side). No CNT bridge was formed with the backside heating although all other conditions were kept the same as the front-side heating, which demonstrated the influence of direct laser irradiation on the CNT assembly. Since both gas flow and van der Waals forces remain unchanged, the absence of CNT bridges is mainly due to the absence of optical gradient forces during backside laser heating. Therefore, the near-field enhanced optical gradient forces should be the dominant reason for the optical assembly of CNT devices.

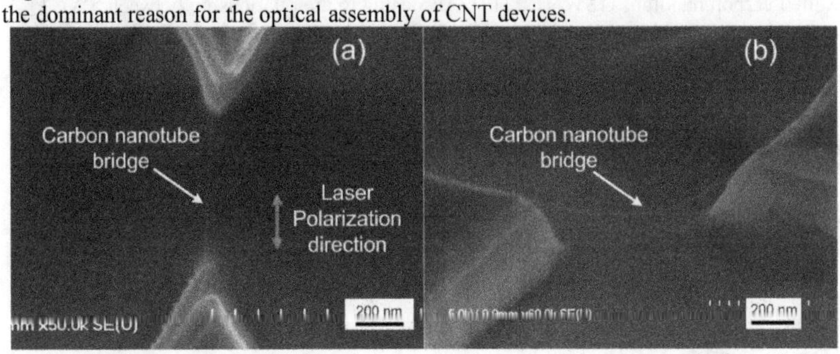

Figure 2. SEM images of a CNT bridge integrated between two sharp electrode tips. (a) Top and (b) side views of the CNT bridge.

Numerical simulations

To obtain theoretical insight of the laser-assisted CVD process, numerical simulations using finite element method have been carried out to simulate the optical near-field effects around tip apexes under CO_2 laser irradiation. Simulation results indicate that a localized heating effect and an optical trapping effect are generated by the laser interaction with sharp tip-shaped electrodes, which suggests the mechanisms of controlled synthesis and assembly of CNTs by optical near-field effects. Figure 3(a) shows that the temperature of tip apexes region is over 200 °C higher than that of the rest regions. Since the CNT synthesis is highly dependent of temperature [16], the localized heating effect can provide an easy way to control the CNT growth location. Moreover, the trapping force distribution as shown in Figure 3(b) suggests that an optical trap is formed between electrode tips under a linearly polarized laser beam. Attractive force field is generated around the tip apexes whose axis is parallel to the laser polarization direction.

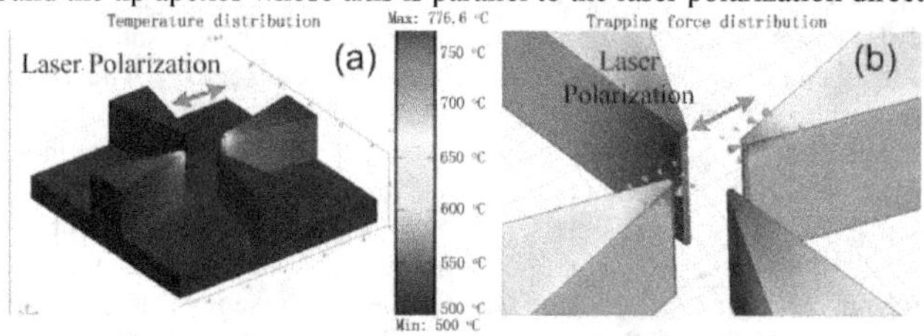

Figure 3. Numerical simulation of optical near-field effects. (a) temperature distribution around electrode tip area. (b) trapping force distribution around electrode tip area. The pink arrows in (b) are scaled to ∇E^2 (gradient of the square of the electric field) magnitudes and indicate the directions of gradient forces acting on CNTs.

Raman analysis and I-V characterization

Figure 4(a) shows a typical Raman spectrum of a CNT bridge using a Renishaw inVia Raman microscope with a 488 nm laser excitation. The sharp radial breathing mode (RBM) peak at 204.8 cm^{-1}, G^+ peak at 1591.5 cm^{-1} and G^- peak at 1565.5 cm^{-1} clearly indicate the existence of single-walled carbon nanotubes (SWNTs) [17]. According to the relationship between SWNT diameter (*dt*) and the radial breathing mode (RBM) frequency (ω_{RBM}) for isolated SWNTs [17], $\omega_{RBM} = 248/dt$, the SWNT diameter is estimated to be 1.2 nm. In the Raman spectrum a weak and wide D peak centered at 1337.5 cm^{-1} is also observed which represents the degree of defects or perhaps amorphous carbon. However, the low ratio of the peak intensity of D-band to G-band indicates that the CNT samples prepared at a low substrate temperature of 500 °C are still in good quality. Figure 4(b) shows the current-voltage (I-V) characteristics of a representative device. The nonlinear I_{DS}-V_{DS} curves and the inverse relationship between I_{DS} and V_G (gate voltage) clearly show the p-type semiconducting characteristics of the CNT between two electrodes.

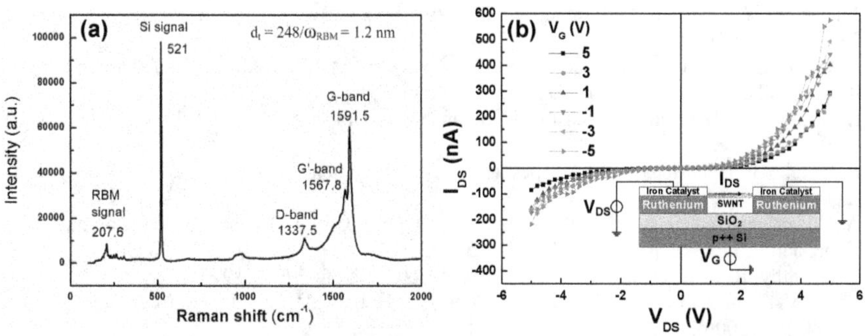

Figure 4. (a) Typical Raman spectrum of the CNT bridges; and (b) Typical I-V characteristics of the CNT devices.

CONCLUSIONS

An optical assembly method was developed for the fabrication of CNT-based FETs using tip-induced optical traps. Controlled integration of individual CNTs bridging predefined electrodes has been achieved with a sub-20-nm accuracy in a LCVD process. The high assembly accuracy is ascribed to the strong confinement of CNTs by tip-induced optical trapping and localized Joule heating effects. This method combines the growth and integration of CNTs into one step, which greatly simplifies the fabrication of CNT devices and avoids the CNT contamination from solution treatment. Moreover, no physical contact to the electrodes is required in this method to introduce the driving force for the CNT assembly, which paves the way to integrate CNTs with large numbers of isolated tiny electrodes for nanoelectronic application.

ACKNOWLEDGMENTS

This research work was financially supported by the National Science Foundation (CMMI 0555884 and ECCS 0621899) and the Nebraska Research Initiative. We thank Prof. N. J. Ianno in the Department of Electrical Engineering, Drs. J. Hua and L. P. Yue in the Department of Physics and Astronomy, and Drs Y. Zhou and H. Chen in Biotechnology Center for their technical support and helpful discussion.

REFERENCES

1. Z. H. Chen, J. Appenzeller, Y. M. Lin, J. Sippel-Oakley, A. G. Rinzler, J. Y. Tang, S. J. Wind, P. M. Solomon, and P. Avouris, *Science* **311**, 1735-1735 (2006).
2. S. Maeng, S. Moon, S. Kim, H. Y. Lee, S. J. Park, J. H. Kwak, K. H. Park, J. Park, Y. Choi, F. Udrea, W. I. Milne, B. Y. Lee, M. Lee, and S. Hong, *Appl. Phys. Lett.* **93**, 113111 (2008).
3. J. G. Nam, Y. J. Park, B. S. Kim, and J. S. Lee, *Scripta. Mater.* **62**, 148-150 (2010).
4. V. Sazonova, Y. Yaish, H. Ustunel, D. Roundy, T. A. Arias, and P. L. McEuen, *Nature* **431**, 284-287 (2004).

5. J. Lefebvre, J. F. Lynch, M. Llaguno, M. Radosavljevic, and A. T. Johnson, *Appl. Phys. Lett.* **75**, 3014-3016 (1999).
6. Y. G. Zhang, A. L. Chang, J. Cao, Q. Wang, W. Kim, Y. M. Li, N. Morris, E. Yenilmez, J. Kong, and H. J. Dai, *Appl. Phys. Lett.* **79**, 3155-3157 (2001).
7. K. Keren, R. S. Berman, E. Buchstab, U. Sivan, and E. Braun, *Science* **302**, 1380-1382 (2003).
8. R. Krupke, F. Hennrich, M. M. Kappes, and H. V. Lohneysen, *Nano Lett.* **4**, 1395-1399 (2004).
9. A. Vijayaraghavan, S. Blatt, D. Weissenberger, M. Oron-Carl, F. Hennrich, D. Gerthsen, H. Hahn, and R. Krupke, *Nano Lett.* **7**, 1556-1560 (2007).
10. L. Novotny, R. X. Bian, and X. S. Xie, *Phys. Rev. Lett.* **79**, 645-648 (1997).
11. J. Shi, Y. F. Lu, K. J. Yi, Y. S. Lin, S. H. Liou, J. B. Hou, and X. W. Wang, *Appl. Phys. Lett.* **89**, 083105 (2006).
12. M. Cantoro, S. Hofmann, S. Pisana, V. Scardaci, A. Parvez, C. Ducati, A. C. Ferrari, A. M. Blackburn, K. Y. Wang, and J. Robertson, *Nano Lett.* **6**, 1107-12 (2006)
13. T. K. Jang, J. H. Ahn, Y. H. Lee, and B. K. Ju, *Chem. Phys. Lett.* **372**, 745-749 (2003).
14. W. Xiong, Y. S. Zhou, M. Mahjouri-Samani, W. Q. Yang, K. J. Yi, X. N. He, S. H. Liou, and Y. F. Lu, *Nanotechnology* **20**, 025601 (2009).
15. Y. S. Zhou, W. Xiong, Y. Gao, M. Mahjouri-Samani, M. Mitchell, L. Jiang, and Y. F. Lu, *Nanotechnology* **21**, 315601 (2010).
16. Y. G. Yao, Q. W. Li, J. Zhang, R. Liu, L. Y. Jiao, Y. T. Zhu, and Z. F. Liu, *Nature Materials* **6**, 283-6 (2007).
17. M. S. Dresselhaus, G. Dresselhaus, R. Saito, and A. Jorio, *Physics Reports-Review Section of Physics Letters* **409**, 47-99 (2005).

Mater. Res. Soc. Symp. Proc. Vol. 1365 © 2011 Materials Research Society
DOI: 10.1557/opl.2011.1386

High Photosensitivity Two-Photon Photoresists for Large Area Surface Microstructuring

Robert J. DeVoe[1], Tzu-Chen Lee[1], Jeremy K. Larsen[1], David A. Ender[1], Jennifer J. Sahlin[1], Craig R. Sykora[2], Cheryl A. Patnaude[1], Matthew R. Atkinson[1], Michael E. Griffin[1], Brian J. Gates[1], and David H. Redinger[1]

[1]Corporate Research Labs, 3M Co., 201-3N-05 3M Center, St. Paul, MN 55144, U.S.A.

[2]Electronic Markets and Materials Lab, 3M Co., St. Paul, MN 55144, U.S.A.

ABSTRACT

Two-Photon initiated polymerization (TPIP) has shown great promise for fabrication of complex micro- and nano-structures. The method has been used to fabricate such structures over small areas (< 1 mm^2) because of slow fabrication speeds and resulting long fabrication times. In order for TPIP to reach practical application in a commercial setting fabrication times need to be reduced by orders of magnitude. We report results on a highly photosensitive initiation system for photoresists based on free radical and cationic polymerization, where photosensitivity is increased 10^2- to 10^3-fold compared to previously reported photoinitiation systems. Threshold writing speeds are determined for critical exposure conditions, including laser power, type and concentration of photoinitiation system, and photoresist type. Surface roughness, a critical parameter in applications such as optics and microfluidics, for example, is also used to determine threshold writing speed. The utility of the approach is demonstrated by making a cell phone keypad light guide from a microreplication tool fabricated using the highly photosensitive photoresist.

INTRODUCTION

Two-photon initiated polymerization (TPIP) has emerged as one of the most flexible methods of fabricating complex micro- and nano-structures. Many optical, mechanical, biological and medical applications can be addressed uniquely by TPIP.[1,2,3] The method has been limited to fabricating arrays of structures over small areas (< 1 mm^2) because of fabrication speeds ≤ 10^3 μm/sec and resulting fabrication times that can reach days or weeks for areas over 1 mm^2.[4] In order for TPIP to reach practical application in a commercial setting fabrication times need to be reduced by orders of magnitude. Most work using TPIP has used a spot-writing method, a method that enables fabrication of very complex structures, but has limited throughput, i.e., limited area that can be fabricated in a reasonable timeframe. Work using multiple beams[5] and dynamic mask exposure[6] has shown some promise in improving throughput.

All of these writing strategies are limited by the low efficiency for initiation of polymerization that is intrinsic to the two-photon process. The rate of initiation of polymerization (v_{ip} in moles/liter/sec) in TPIP is proportional to the square of the light intensity (I^2 in watts/cm^2), the cross-section for simultaneous absorption of two photons (δ in cm^4 sec photon^{-1}), the concentration of the light-absorbing molecule ([Dye], moles/liter), and the quantum efficiency of initiation of polymerization (σ, no units):[2]

$$\nu_{ip} \ \alpha \ I^2\delta[\mathrm{Dye}]\sigma \qquad (1)$$

Many reports have appeared describing dyes with large two-photon absorption cross-sections.[2,7] Commercial photoinitiators have high efficiencies for initiation of polymerization when exposed under conditions for single photon absorption. Few photoinitiation systems have been reported that provide both high efficiency for initiation of polymerization and large two-photon absorption cross-sections.[3] Such systems are critical to increasing the throughput the orders of magnitude required for TPIP to reach practical application in a commercial setting.

We describe a high photosensitivity photoinitiation system and the relationships between writing speeds and critical imaging parameters, including cross-hatch density, slice thickness, and photoresist type. The utility of the photoinitiation system is shown by fabricating a microreplication tool ~20 cm^2 in size and using the tool to fabricate a cell phone keypad light guide.

EXPERIMENTAL

Photoresists

Acrylate photoresists containing photoinitiation system, 30% poly(methyl methacrylate) (120,000 MW, Aldrich), 35% SR9008 (trifunctional acrylate, Sartomer), and 35% SR368 (trifunctional acrylate, Sartomer), were prepared as a 50% solids solution in cyclopentanone. After filtering through a 0.25 micron PTFE syringe filter the photoresist was coated on silicon wafers primed with 3-triethoxysilylpropyl methacrylate by spin coating, resulting in dry film thickness of about 20 microns. SU8 photoresist (MicroChem) was prepared with photoinitiation system, filtered through a 0.25 micron PTFE syringe filter, and coated on silicon wafers primed with 2-(3,4-epoxycyclohexyl)ethyltrimethoxysilane by spin coating, resulting in dry film thickness of about 20 microns. Post-exposure bake of 3.5 minutes at 90 C was used with SU8.

KL68 was synthesized as reported.[8] Iodonium salts were obtained from Alpha Aesar (Ph_2I^+ (DPI) PF_6^-) and Hampford Research (bis-(4-t-butylphenyl)iodonium SbF_6^-). Alkyl borate (CGI7460) and Irgacure 369 were obtained from Ciba Specialties (Figure 1).

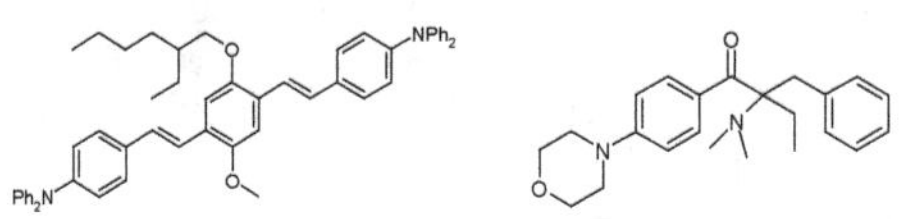

KL68 Irgacure 369

Figure 1. Structures of photosensitizer KL68 and photoinitiator Irgacure 369.

Exposures

For all samples, exposures were carried out using a two-photon writing system with a Spectra-Physics Ti-sapphire laser (800 nm center wavelength, about 80 fs pulse width, and 80

MHz repetition rate), an electro-optic shutter, power control, dispersion compensation, confocal interface detection, a custom high speed galvo scanner (galvos from Cambridge Technologies, Lexington, MA), and an oil immersion objective (40x, NA 1.0) with a focus waist of 580 nm (Figure 2).[9] Sample positioning is controlled using a Newport XY air bearing stage. Stage and galvo scanner are controlled by a custom CAD-based control system. This system is capable of scan speeds of about 1 to >100,000 microns per second and large writing area (up to 600 cm^2).

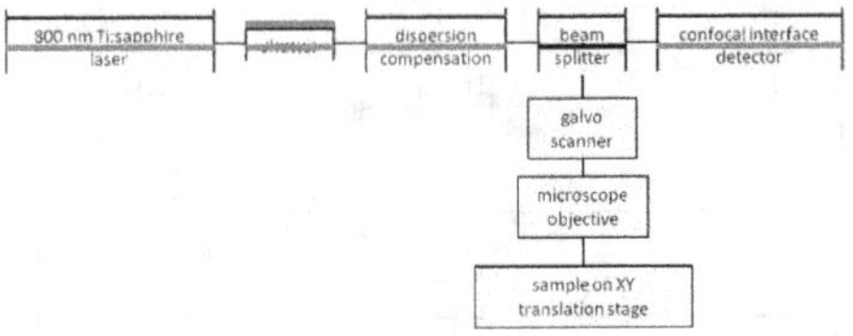

Figure 2. Schematic of exposure apparatus.

Analysis

The surface topography for each test structure was measured with a Wyko NT900 optical interferometer, without coating the samples (50x objective, 0.5x field lens, 2% modulation threshold). It was observed that a reflection of the top surface of the feature (reflected from the Si wafer surface) appeared when scanning the sample through focus. Care was taken so that this reflection was not included in the scan data; this was easiest for features greater than 3 microns tall. Some features were too short for optical interferometry (as the vertical position of the internal reflection of the top of the feature was near level with the Si surface outside of the feature). These features were measured via confocal microscopy (Keyence VK9700 confocal microscope, 50x/0.95 objective).

The topographic maps were analyzed using a routine written in Matlab. This routine identified the central portion of the top of a feature and calculated the mean height of the feature. A spherical surface was fit to the top surface of the feature, and the radius of curvature calculated. Curvature and tilt were removed from the top surface topography and the rms roughness (Rq) was calculated.

RESULTS & DISCUSSION

Threshold writing speeds were determined for photoinitiation systems by writing arrays of lines where each line was written $\sqrt{2}$ times faster than the previous line, in accordance standard imaging science convention.[10] Threshold writing speed was taken as the highest scan speed where a line was observed. Photosensitivities relative to Irgacure 369 were 45 (KL68, acrylate), 128 (KL68, DPI PF6, and CGI7460, acrylate), and >180 (KL68, DPI SbF6, SU8 epoxy), with threshold writing speeds in excess of 100,000 μm/s for the latter two. Irgacure 369 was used at 1% by weight, KL68 at 0.5%, DPI salts at 1%, and CGI7460 at 0.5%.

Surface roughness was measured as a function of writing parameters and photoinitiation system. Boxes 50 x 50 x 10 micron (l x w x h) were written and surface roughness measured by optical interferometry. Arrays of boxes were written at different scan speeds, cross-hatch spacing, and slice thickness, for each of the combinations of photoresist type and photoinitiation system. Table I shows representative data on surface roughness as a function of scan speed, cross-hatch spacing, and slice thickness, for acrylate photoresist with KL68, KL68 + DPI PF6, and KL68 + DPI PF6 + CGI7460.

Table I. Surface roughness, R_q, of boxes fabricated with parameters in table using photoinitiation system 0.5% KL68 + 0.5% DPI PF6 + 1.0% CGI 7460 in acrylate photoresist, 7 mW, 80 MHz, 580 nm spot size.

Scan speed →	25.6k μm/s	51.2k μm/s	72.4k μm/s	102.4k μm/s
0.25 μm hatch, 0.5 μm slice	8 nm	10.6 nm	13.4 nm	39 nm
0.5 μm hatch, 0.5 μm slice	11.7 nm	33.7 nm	39.8 nm	44.8 nm
1.0 μm hatch, 0.5 μm slice	69.5 nm	159 nm	157 nm	144 nm

Large area microfabrication was demonstrated using the high-photosensitivity photoresists to fabricate a master tool for a cell phone keypad light guide. In commercial cell phones keys are generally lit by a light guide with light extraction regions directly beneath the key graphics to minimize wasted light, and one or more LEDs. Design and fabrication of a high-efficiency light guide minimizes the required number of LEDs, reducing complexity and increasing battery lifetime. We have chosen to create physical structures penetrating into the light guide to reflect the light in desired directions.

An off-the-shelf cell phone was purchased and deconstructed. The keypad light guide used a total of ten LEDs to illuminate the keys. A new light guide was designed, fabricated, and inserted into the cell phone shell, requiring only one LED to achieve the desired key brightness and uniformity. The master tool, 5.48 x 3.35 cm (18.4 cm^2), was fabricated in 47.5 hours, using a scan rate of 40,000 μm/s, hatch spacing of 1 μm, and slice thickness of 2 μm, laser power 7.5 mW, 80 MHz, 580 nm spot diameter.

The basic extractor shape chosen was a paraboloid, varying base radius, height, and truncation of the top. Examples are given in Figure 3. A triangular deflector was located directly in front of the LED, its height being less than the light guide thickness of 0.3 mm to deflect only a portion of the light, with the remainder passing above it.

Figure 3. (a) Extractor height 72 μm, base radius 60 μm, (b)Same extractor parameters as (a) but truncated 37 μm from the base, (c) Extractor height 18 μm, base radius 30 μm.

Optimization of extractor parameters and density was performed using ASAP ray tracing software (from Breault Research Organization) and Isight optimization software (from Dassault Systèmes). Some of the keys were divided into two regions to achieve optical uniformity and eliminate hot spots. The final design had 33,031 extractors consisting of eight distinct shapes and one deflector. Figure 4 shows false color luminance of the simulated light guide along with photos of the fabricated light guide illuminating the keypad.

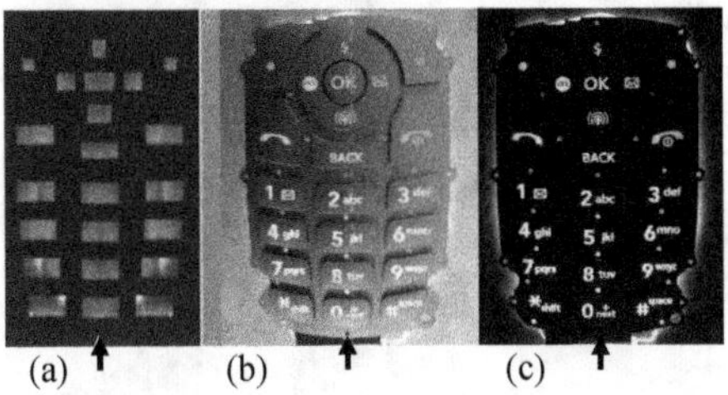

(a) (b) (c)

Figure 4. (a) Simulated light guide extraction, (b) photo of keypad illuminated by fabricated light guide with room lights on, and (c) with room lights off. Arrow shows LED location.

CONCLUSIONS

We have shown that high sensitivity two-photon photoresists based on two-photon sensitization of iodonium salts and borate salts enable fabrication of master tools of commercially meaningful size in a reasonable timeframe. These photoresists were used to fabricate master tools of light extractor patterns for a cell phone keypad light guide that enables backlighting with a single LED with high, uniform light output.

REFERENCES

1. Y.-L. Zhang, Q.-D. Chen, H. Xia, H.-B. Sun, *Nano Today* **5**(5), 435-448 (2010).
2. M. Rumi, S. Barlow, J. Wang, J. W. Perry, S. R. Marder, *Adv. Polym. Sci.,* Springer-Verlag (2008), vol 213:1-95
3. K. D. Belfield, S. Yao, M. V. Bondar *Adv. Polym. Sci*, Springer-Verlag (2008), vol 213:97-156.
4. M. P. Stocker, L. Li, R. R. Gattass, J. T. Fourkas, *Nature Chem.* **3**, 223-227 (2011); T. S. Hsieh, C. W. B. Ng, K. Narayanan, A. C. A. Wan, J. Ying, *Biomater.* **31**, 7648-7652 (2010); K. Obata, J. Koch, U. Hinze, B. N. Chichkov, *Opt. Express* **18**, 17193-17200 (2010).
5. J. Kato, N. Takeyasu, Y. Adachi, H.-B. Sun, S. Kawata *Appl. Phys. Lett.* **86**(4), 044102 (2005); P. R. Fleming, R. J DeVoe, N. A. Stacey, C. A. Leatherdale, R. D. DeMaster, T. A. Ballen, J. M Florczak, U.S. Patent Appl. Publ. US 2004/0012872 (22 January 2004).
6. R. Nielson, B. Kaehr, J. B. Shear *Small* **1**(1), 120-125 (2009).
7. B. A. Reinhardt, L. L. Brott, S. J. Clarson, A. G. Dillard, J. C. Bhatt, R. Kannan, L. Yuan, G. S. He, P. N. Prasad *Chem. Mater.* **10**, 1863-1874 (1998).
8. C. A. Leatherdale, C. R. Schardt, D. S. Thompson, W. L. Thompson, U.S. Patent No. 7 265 161 (4 September 2007).
9. For more detailed description of the exposure apparatus see C. M. Marttila, C. D. Hoyle, C. A. Ender, J. B. Willoughby, R. J. DeVoe, U.S. Patent No. 7 941 013 (10 May 2011).
10. A. Reiser, Photoreactive Polymers: Science and Technology of Resists, J. Wiley and Sons, (1989).

Mater. Res. Soc. Symp. Proc. Vol. 1365 © 2011 Materials Research Society
DOI: 10.1557/opl.2011.1037

Optimization of Cavity Length to Obtain Low Threshold Solid-state Dye Laser

Lan-Ting Shi,[1] Feng Jin,[1] Mei-Ling Zheng,[1] Xian-Zi Dong,[1] Wei-Qiang Chen,[1] Zhen-Sheng Zhao,[1] and Xuan-Ming Duan[1,*]
[1]Laboratory of Organic NanoPhotonics and Key Laboratory of Functional Crystals and Laser Technology, Technical Institute of Physics and Chemistry, Chinese Academy of Sciences, Zhongguancunbeiyitiao No. 2, Beijing, 100190, People's Republic of China

ABSTRACT

We have demonstrated and studied polymeric solid-state dye lasers (SSDLs) fabricated by three-dimensional (3D) polystyrene colloidal crystals and *tert*-butyl roadamine B (*t*-Bu RhB) doped Poly (methyl methacrylate) (PMMA) films with different film thickness. The sandwich-typed resonator cavities with different active layer thickness display single-mode lasing oscillations in the reflection bandgap of the colloidal crystals. The lasing thresholds could be optimized by changing the thickness of *t*-Bu RhB doped PMMA films, which is as low as 7.43 W/cm^2. Adjusting active layer thickness would provide an opportunity to accelerate the development of fabricating polymeric SSDLs with low threshold.

INTRODUCTION

Polymeric solid-state dye lasers show great promise for the advantages of compact, low-cost, tunable, hazard-less, easy to fabrication, efficient laser sources and have been expected to be applied in the numerous fields [1-3]. Polystyrene (PS) colloidal crystals (CCs) have been reported to construct Fabry-Perot cavity as resonator materials for obtaining lasing oscillations [4]. Basically, the SSDLs are constructed with a resonator cavity to provide optical feedback and an active medium to acquire high optical gain. In a Fabry-Perot laser cavity, the thickness of active layer is corresponding to the cavity length, which is a significant parameter to fabricate high-efficient low threshold laser [5]. Consequently, lasing threshold is expected be optimized by changing the thickness of active layer, which is significant for the future application of the polymer-based SSDLs. However, optimization of the lasing threshold in such resonator cavity was seldom reported. In the present study, we demonstrated optimizing lasing thresholds by adjusting the gain medium thickness in the Fabry-Perot resonator cavity.

EXPERIMENTAL DETAILS

The PS CCs were obtained by the following procedures [6]. Firstly, monodisperse PS microspheres were synthesized by emulsion polymerization of styrene, deionized water, surfactant sodium dodecyl sulfonate, buffer agent of ammonium bicarbonate and KPS-water solution. Then, PS colloidal crystals were deposited on glass substrate by vertical deposition technique at 65℃ for 72 hours. And PS CC on one side of substrate was scraped off, while the other side is left for characterization and cavity fabrication.

The dye-doped polymer film was prepared by incorporating *tert*-butyl roadamine B (*t*-Bu RhB, with molecular weight, MW=512 g/mol) into PMMA photoresist, as in our previous work [7]. The concentration of *t*-Bu RhB was 3.0 wt%. After irradiation of the dye-doped photoresist under a high-voltage mercury lamp for 3 minutes, a transparent film was formed on glass substrate. The liquid photoresist was put between two cover glasses sandwiched by a rubber spacer and thus *t*-Bu RhB-doped films with different thicknesses were prepared be controlled by tuning the spacer thickness.

The resonator cavities were prepared by employing the *t*-Bu RhB doped PMMA films with different thickness and pairs of PS CCs, as in our previous report [8]. The resonator cavity appeared as a "sandwich", as shown in Figure 1.

In order to measure the emission spectra from the resonator cavities, the optical measuring setup was presented in Figure 1. The pumping source was a Q-switched Nd:YAG laser operating at the wavelength of 532 nm, having a pulse duration of 8 ns and a repetition frequency of 10 Hz. The laser beam was focused on the sample using a lens. The incident angle was about 45°oblique to (111) plane of the PS CCs. Incident intensity was controlled by a neutral-density filter and measured by a power meter. The optical emission spectra were measured from the opposite side of the excitation beam by using a fiber-coupled grating spectrometer (USB2000, Ocean Optics) and detection direction was vertical to the sample. All the measurements were performed at room temperature.

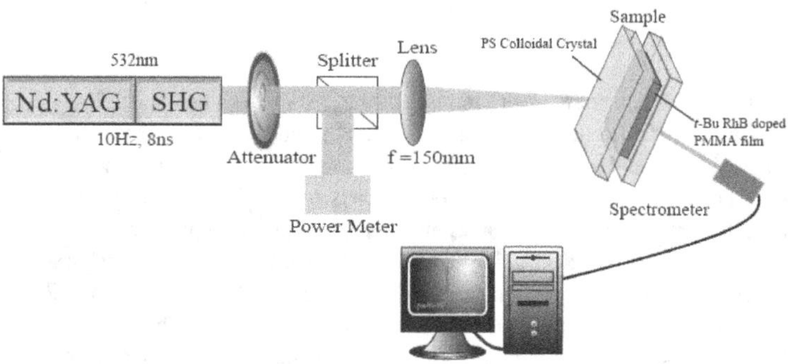

Figure 1. Schematic illustration of the sandwich-typed resonator cavity and optical system for measurement.

RESULTS AND DISCUSSION

Inspections of the PS CCs were carried out by scanning electron microscopy (SEM), as shown in Figure 2 (a). The PS CCs were of a good structural quality. The PS nanoparticle paced in a face-centered-cubic (fcc) lattice oriented in the (111) direction on the top surface of the substrate. The transmission spectrum of PS CCs was measured by the Shimadzu UV-2550 spectrometer, as shown in Figure 2 (b). The photonic bandgap (PBG) of the PS CCs was centered at 596 nm. The inset of Figure 2 (b) was the molecular structure of the *t*-Bu RhB, which was doped in the PMMA photoresist. And we also characterized the optical property of the *t*-Bu RhB doped PMMA film, taking the sample with film thickness of 7.3 μm as an example. The emission spectrum of the film was presented in Figure 2, which exhibited the photoluminescence (PL) with the maximum intensity at the wavelength at 596 nm. The PBG of PS CCs overlapped with the photoluminescence band of the *t*-Bu RhB doped PMMA film. Hence, it is predictable that lasing action could be efficiently generated by the internal feedback effect of the cavity under the irradiation of the pumping source.

In order to achieve low-threshold resonator cavity, we have prepared *t*-Bu RhB doped PMMA films with different thickness. With changing the stress on the glass substrate, the *t*-Bu RhB doped PMMA films with thickness of 7.3, 9.4, 11.3 μm were obtained. Figure 3 showed the SEM images of *t*-Bu RhB doped PMMA films, which exhibited the cross-section of the film.

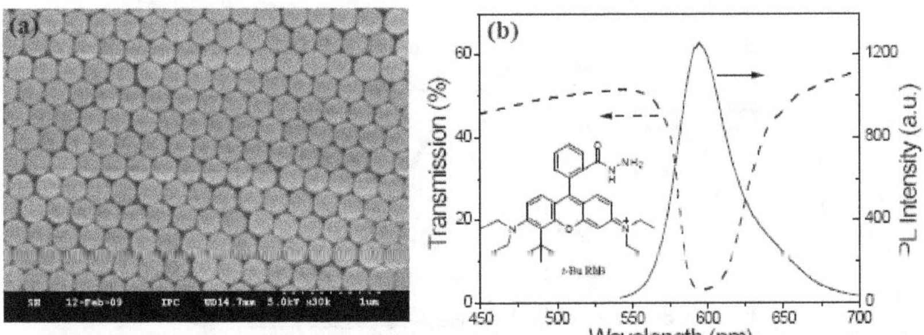

Figure 2. (a) SEM image of the PS CC from the top view. (b) Transmission spectrum of the PS CCs (dash line); photoluminescence spectrum of the *t*-Bu RhB doped PMMA film with film thickness of 7.3 μm (solid line); the inset is the molecular structure of the *t*-Bu RhB.

Figure 3. The SEM images of the *t*-Bu RhB doped PMMA films with thickness of 7.3, 9.4, 11.3 μm.

Emission spectra of resonator cavities with different thickness *t*-Bu RhB doped PMMA films were measured using the setup presented in Figure 1. Figure 4 (a) showed the emission spectra of resonator cavity with 11.3 μm *t*-Bu RhB doped PMMA film both below and above lasing threshold. There is no obviously emission peak appeared in the emission spectra at low excitation intensity (below lasing threshold). When the excitation energy exceeds the lasing threshold, a narrow single mode emission occurs at near 592 nm, which is corresponding to the PBG of PS PCs. The full-width half maximum (FWHM) of lasing peaks is 0.70 nm. In order to determine the threshold excitation energy for 11.3 μm thickness sample, we evaluated the dependence of the emission intensity and FWHM on the excitation energy. Fig.4 (b) presents the changes in the emission intensity and FWHM at the laser wavelength of 592 nm for the resonator cavity with 11.3 μm thickness film as a function of the excitation energy. As is evident from this figure, the lasing threshold is determined from the intersection of the two line fits to the intensity below and above threshold, is 11.97 W/cm^2.

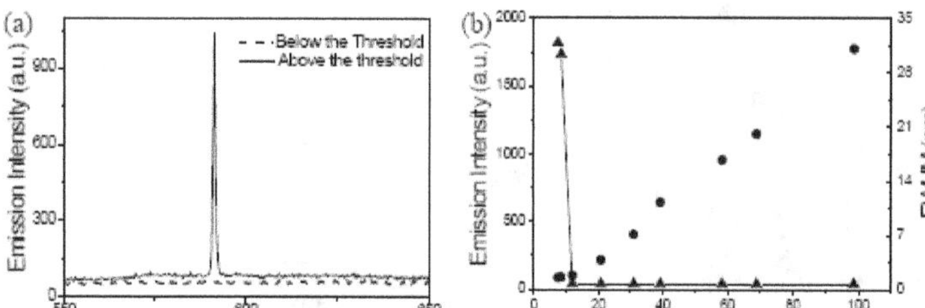

Figure 4. (a) Emission spectra of the resonator cavity with film thickness of 11.3 μm below and above lasing threshold. (b) Changes in the emission intensity (filled circle) and FWHM (filled triangle) for resonator cavity as a function of excitation energy.

The emission spectra of resonator cavities with film thickness of 7.3 and 9.4 μm were measured in the same way as the sample with film thickness of 11.3 μm. Lasing emission of resonator cavities with *t*-Bu RhB doped PMMA films thickness of 7.3 and 9.4 μm were all single-mode and centered at 592nm. The spectral width (full-width half maximum, FWHM) of lasing peaks were 0.71 and 0.70 nm. The lasing thresholds as functions of thickness for *t*-Bu RhB doped PMMA films were presented in Figure 5. The results in Figure 5 demonstrated that there was an optimum active layer thickness, at which the resonator cavity exhibited a minimum lasing threshold. These results suggested that the lasing threshold of sandwich-typed resonator cavity could be optimized by changing the thickness of the dye doped active layer. Therefore, we could obtain high-efficient polymeric SSDL with low lasing threshold through optimizing the parameter of active layer thickness. The optimization of active layer has huge advantage in fabricating low threshold and high performance solid-state dye lasers.

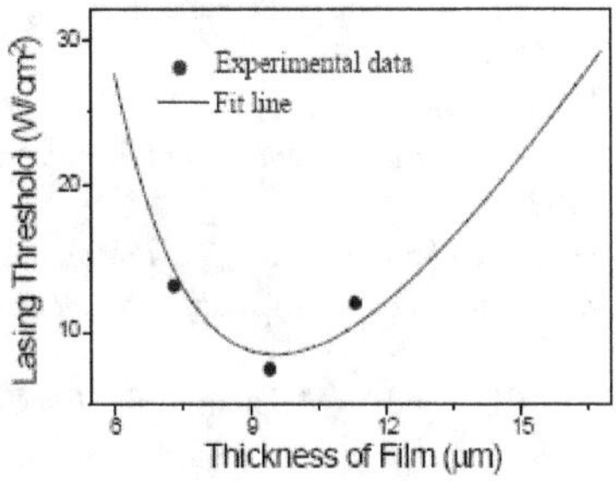

Figure .5 Lasing threshold changing with thickness of *t*-Bu RhB doped PMMA film.

CONCLUSIONS

In summary, we demonstrated polymeric sandwiched resonator cavity fabricated by *t*-Bu RhB doped polymer film and colloidal crystals. Low threshold and single mode laser oscillations were observed from laser cavities with *t*-Bu RhB doped polymer films of different thicknesses. The lasing threshold has been optimized by adjusting the thickness of *t*-Bu RhB doped polymer film. As a result, changing the thickness of dye-doped polymer film in the resonator cavity would provide a simple and convenient method to optimize the lasing threshold of optically pumped polymeric solid-state dye laser.

ACKNOWLEDGMENTS

The authors are grateful to the National Natural Science Foundation of China (Grant No. 50773091, 50973126, 60907019, 51003113), the National Basic Research Program of China (2010CB934103), International Cooperation Program of MOST (2008DFA02050, 2010DFA01180) for financial supports.

REFERENCES

1. R. Sastre and A. Costela, *Adv. Mater.* **7,** 198 (1995).
2. A. Costela, I. Garcia-Moreno and R. Sastre, *Phys. Chem. Chem. Phys.* **5**, 4745 (2003).
3. P. Del Carro, A. Camposeo, R. Stabile, E. Mele, L. Persano, R. Cingolani and D. Pisignano, *Appl. Phys. Lett.* **89**, 201105 (2006).
4. H. Yamada, T. Nakamura, Y. Yamada and K. Yano, *Adv. Mater.* **21**, 4134 (2009).
5. C. C. Davis, "*Lasers and Electro-Optics: Fundamentals and Engineering*" (Cambridge University Press, 1996) pp. 88-107.
6. P. Jiang, J. F. Bertone, K. S. Hwang and V. L. Colvin, *Chem. Mater.* **11**, 2132 (1999).
7. L. T. Shi, F. Jin, W. Q. Chen, Z. S. Zhao and X. M. Duan, *Proc. of SPIE* **7774**, 77740V-1 (2010).
8. F. Jin, C. F. Li, X. Z. Dong, W. Q. Chen and X. M. Duan, *Appl. Phys. Lett.* **89**, 241101 (2006).

AUTHOR INDEX

SUBJECT INDEX

For EU product safety concerns, contact us at Calle de José Abascal, 56–1°,
28003 Madrid, Spain or eugpsr@cambridge.org.

www.ingramcontent.com/pod-product-compliance
Lightning Source LLC
LaVergne TN
LVHW011519200926
848484LV00001B/77